MATHEMATICS POW G
FOR CHILDREN

WORKBOOK ONE

Everard Barrett

CONTEXTUAL MATHEMATICS TEACHING METHODOLOGY

Professor B Enterprises, Inc.
P.O. Box 3856
Suwanee, GA 30024
www.profb.com

FOURTH EDITION REVISED

Copyrights and acknowledgements

Editor
Everard Barrett

Production Manager
Veda Barrett

Cover Art
Charles J. Berger

Publisher
Professor B Enterprises, Inc.
P. O. Box 3856
Suwanee, GA 30024
www.profb.com

Published in the United States by Professor B Enterprises, Inc. Portions of this workbook were previously published under the following titles:

FOURTH EDITION REVISED

ISBN 1-883324-04-1

Printed in the United States of America

Table of Contents

		PAGE
FACILITY EXERCISES **#1**	UNDERSTANDING "ONE" THROUGH "TEN"	1
FACILITY EXERCISES **#2**	LEARNING TO COUNT FROM ONE TO TEN	2
FACILITY EXERCISES **#3**	COUNTING BY TWOS, THREES, FOURS	3
FACILITY EXERCISES **#4**	MORE, LESS, EQUAL	4
FACILITY EXERCISES **#5**	ADDITION FACTS	5
FACILITY EXERCISES **#6**	LINKING ADDITION AND SUBTRACTION	7
FACILITY EXERCISES **#7**	SUBTRACTION FACTS	9
FACILITY EXERCISES **#8**	ADDING AND SUBTRACTING WITH ZERO	12
FACILITY EXERCISES **#9**	SUMS AND DIFFERENCES INVOLVING MORE THAN TWO ADDENDS	13
FACILITY EXERCISES **#10**	THE LOWER ADDITION AND SUBTRACTION FACTS	15
FACILITY EXERCISES **#11**	THE LOWER ADDITION AND SUBTRACTION FACTS	15
FACILITY EXERCISES **#12**	THE LOWER ADDITION AND SUBTRACTION FACTS	16
FACILITY EXERCISES **#13**	THE LOWER ADDITION AND SUBTRACTION FACTS	16
FACILITY EXERCISES **#14**	THE LOWER ADDITION AND SUBTRACTION FACTS	17
FACILITY EXERCISES **#15**	THE LOWER ADDITION AND SUBTRACTION FACTS	17
FACILITY EXERCISES **#16**	THE LOWER ADDITION AND SUBTRACTION FACTS	18
FACILITY EXERCISES **#17**	THE LOWER ADDITION AND SUBTRACTION FACTS	18
FACILITY EXERCISES **#18**	THE LOWER ADDITION AND SUBTRACTION FACTS	19
FACILITY EXERCISES **#19**	THE LOWER ADDITION AND SUBTRACTION FACTS	20
FACILITY EXERCISES **#20**	THE LOWER ADDITION AND SUBTRACTION FACTS	21
FACILITY EXERCISES **#21**	THE LOWER ADDITION AND SUBTRACTION FACTS	22
FACILITY EXERCISES **#22**	THE LOWER ADDITION AND SUBTRACTION FACTS	23
FACILITY EXERCISES **#23**	THE LOWER ADDITION AND SUBTRACTION FACTS	24
FACILITY EXERCISES **#24**	THE LOWER ADDITION AND SUBTRACTION FACTS	25
FACILITY EXERCISES **#25**	THE LOWER ADDITION AND SUBTRACTION FACTS	26
FACILITY EXERCISES **#26**	THE LOWER ADDITION AND SUBTRACTION FACTS	27
FACILITY EXERCISES **#27**	THE LOWER ADDITION AND SUBTRACTION FACTS	28
FACILITY EXERCISES **#28**	THE LOWER ADDITION AND SUBTRACTION FACTS	29
FACILITY EXERCISES **#29**	THE LOWER ADDITION AND SUBTRACTION FACTS	30
FACILITY EXERCISES **#30**	THE LOWER ADDITION AND SUBTRACTION FACTS	31
FACILITY EXERCISES **#31**	THE LOWER ADDITION AND SUBTRACTION FACTS	32
FACILITY EXERCISES **#32**	THE LOWER ADDITION AND SUBTRACTION FACTS	33
FACILITY EXERCISES **#33**	THE LOWER ADDITION AND SUBTRACTION FACTS	34
FACILITY EXERCISES **#34**	THE LOWER ADDITION AND SUBTRACTION FACTS	35
FACILITY EXERCISES **#35**	THE LOWER ADDITION AND SUBTRACTION FACTS	36
FACILITY EXERCISES **#36**	THE LOWER ADDITION AND SUBTRACTION FACTS	37
FACILITY EXERCISES **#37**	THE LOWER ADDITION AND SUBTRACTION FACTS	38
FACILITY EXERCISES **#38**	THE LOWER ADDITION AND SUBTRACTION FACTS	39
FACILITY EXERCISES **#39**	THE LOWER ADDITION AND SUBTRACTION FACTS	40
FACILITY EXERCISES **#40**	THE LOWER ADDITION AND SUBTRACTION FACTS	41
FACILITY EXERCISES **#41**	THE LOWER ADDITION AND SUBTRACTION FACTS	42
FACILITY EXERCISES **#42**	THE LOWER ADDITION AND SUBTRACTION FACTS	43
FACILITY EXERCISES **#43**	THE LOWER ADDITION AND SUBTRACTION FACTS	44
FACILITY EXERCISES **#44**	THE LOWER ADDITION AND SUBTRACTION FACTS	45
FACILITY EXERCISES **#45**	THE LOWER ADDITION AND SUBTRACTION FACTS	46
FACILITY EXERCISES **#46**	THE LOWER ADDITION AND SUBTRACTION FACTS	47
FACILITY EXERCISES **#47**	THE LOWER ADDITION AND SUBTRACTION FACTS	48
FACILITY EXERCISES **#48**	THE LOWER ADDITION AND SUBTRACTION FACTS	49
FACILITY EXERCISES **#49**	THE LOWER ADDITION AND SUBTRACTION FACTS	50
FACILITY EXERCISES **#50**	THE LOWER ADDITION AND SUBTRACTION FACTS	51
FACILITY EXERCISES **#51**	THE LOWER ADDITION AND SUBTRACTION FACTS	52
FACILITY EXERCISES **#52**	THE LOWER ADDITION AND SUBTRACTION FACTS	53
FACILITY EXERCISES **#53**	THE LOWER ADDITION AND SUBTRACTION FACTS	54

Table of Contents (continued)

		PAGE
FACILITY EXERCISES #54	THE LOWER ADDITION AND SUBTRACTION FACTS	55
FACILITY EXERCISES #55	THE LOWER ADDITION AND SUBTRACTION FACTS	56
FACILITY EXERCISES #56	THE LOWER ADDITION AND SUBTRACTION FACTS	57
FACILITY EXERCISES #57	THE LOWER ADDITION AND SUBTRACTION FACTS	58
FACILITY EXERCISES #58	COUNTING TO 100 AND BEYOND	59
FACILITY EXERCISES #59	COUNTING TO 100 AND BEYOND	60
FACILITY EXERCISES #60	COUNTING TO 100 AND BEYOND	61
FACILITY EXERCISES #61	VOCABULARY AND SYMBOLISM OF NUMERATION	62
FACILITY EXERCISES #62	VOCABULARY AND SYMBOLISM OF NUMERATION	63
FACILITY EXERCISES #63	VOCABULARY AND SYMBOLISM OF NUMERATION	64
FACILITY EXERCISES #64	VOCABULARY AND SYMBOLISM OF NUMERATION	65
FACILITY EXERCISES #65	VOCABULARY AND SYMBOLISM OF NUMERATION	65
FACILITY EXERCISES #66	VOCABULARY AND SYMBOLISM OF NUMERATION	66
FACILITY EXERCISES #67	ADDING TWO LARGE WHOLE NUMBERS	67
FACILITY EXERCISES #68	PLACE VALUE	70
FACILITY EXERCISES #69	PLACE VALUE	71
FACILITY EXERCISES #70	COMPARING WHOLE NUMBERS	72
FACILITY EXERCISES #71	COMPARING WHOLE NUMBERS	73
FACILITY EXERCISES #72	SUBTRACTING ONE LARGE WHOLE NUMBER FROM ANOTHER	74
FACILITY EXERCISES #73	MIXED PRACTICE	79
FACILITY EXERCISES #74	MIXED PRACTICE	82
FACILITY EXERCISES #75	MIXED PRACTICE	85
FACILITY EXERCISES #76	MIXED PRACTICE	87
FACILITY EXERCISES #77	EQUIVALENT NUMERALS	89
FACILITY EXERCISES #78	EQUIVALENT NUMERALS	90
FACILITY EXERCISES #79	THE TEN–PLUS ADDITION FACTS	91
FACILITY EXERCISES #80	KNOWING "NINE–PLUS," "EIGHT-PLUS," "SEVEN–PLUS," "SIX–PLUS," ADDITION FACTS WITHOUT MEMORIZATION	92
FACILITY EXERCISES #81	KNOWING "NINE–PLUS," "EIGHT-PLUS," "SEVEN–PLUS," "SIX–PLUS," ADDITION FACTS WITHOUT MEMORIZATION	94
FACILITY EXERCISES #82	ADDING ANY TWO WHOLE NUMBERS	95
FACILITY EXERCISES #83	ADDING ANY TWO WHOLE NUMBERS	97
FACILITY EXERCISES #84	EQUIVALENT NUMERALS	100
FACILITY EXERCISES #85	SUBTRACTION FACTS WITHOUT MEMORIZATION	101
FACILITY EXERCISES #86	SUBTRACTING ONE WHOLE NUMBER FROM ANOTHER	105
FACILITY EXERCISES #87	SUBTRACTING ONE WHOLE NUMBER FROM ANOTHER	107
FACILITY EXERCISES #88	SUBTRACTING ONE WHOLE NUMBER FROM ANOTHER	111
FACILITY EXERCISES #89	PREPARING TO "TELL THE TRUTH" WHEN ADDING OR SUBTRACTING WHOLE NUMBERS	115
FACILITY EXERCISES #90	PREPARING TO "TELL THE TRUTH" WHEN ADDING OR SUBTRACTING WHOLE NUMBERS	116
FACILITY EXERCISES #91	"TELLING THE TRUTH" WHEN ADDING WHOLE NUMBERS	117
FACILITY EXERCISES #92	SOLVING WORD PROBLEMS	118
FACILITY EXERCISES #93	"TELLING THE TRUTH" WHEN SUBTRACTING WHOLE NUMBERS	120
FACILITY EXERCISES #94	MIXED PRACTICE	121
FACILITY EXERCISES #95	MIXED PRACTICE	123
FACILITY EXERCISES #96	MIXED PRACTICE	125
FACILITY EXERCISES #97	MIXED PRACTICE	127
FACILITY EXERCISES #98	SOLVING WORD PROBLEMS	129
FACILITY EXERCISES #99	FINDING THE MISSING NUMBER	131
FACILITY EXERCISES #100	FINDING THE MISSING NUMBER	135
FACILITY EXERCISES #101	FINDING THE MISSING NUMBER	136
FACILITY EXERCISES #102	COLUMN ADDITION	137
FACILITY EXERCISES #103	COLUMN ADDITION	139
FACILITY EXERCISES #104	MIXED PRACTICE	141

UNDERSTANDING "ONE" THROUGH "TEN"
FACILITY EXERCISES #1

Do the following **without counting**:

1. Show six fingers.

   Keep showing different sixes.

2. Show two fingers.

   Keep showing different twos.

3. Show nine fingers.

   Keep showing different nines.

4. Show three fingers.

   Keep showing different threes.

5. Show eight fingers.

   Keep showing different eights.

6. Show four fingers.

   Keep showing different fours.

7. Show one finger.

   Keep showing different ones.

8. Show seven fingers.

   Keep showing different sevens.

9. Show five fingers.

   Keep showing different fives.

10. Show ten fingers.

LEARNING TO COUNT FROM ONE TO TEN
FACILITY EXERCISES #2

Do the following:

1. Count forward from one to three.
2. Count backward from three to one.
3. Count forward from one to five.
4. Count backward from five to one.
5. Count forward from one to six.
6. Count backward from six to one.
7. Count from one to eight.
8. Count backward from eight to one.
9. Count forward from one to ten.
10. Count backward from ten to one.
11. Count from two to seven.
12. Count backward from seven to two.
13. Count forward from three to nine.
14. Count backward from nine to three.
15. Count from four to ten.
16. Count backward from ten to four.
17. Count from five to nine and back.
18. Count from three to eight and back.
19. Count from seven to ten and back.
20. Count from one to ten and back.

COUNTING BY TWOS, THREES, FOURS
FACILITY EXERCISES #3

Do the following:

1. Starting at two, count by twos to ten: forward and backward.
2. Starting at one, count by twos to nine: forward and backward.
3. Starting at three, count by threes to nine: forward and backward.
4. Starting at one, count by threes to ten: forward and backward.
5. Starting at two, count by threes to eight: forward and backward.
6. Starting at four, count by fours to eight: forward and backward.
7. Starting at one, count by fours to nine: forward and backward.
8. Starting at two, count by fours to ten: forward and backward.
9. Starting at three, count by fours to seven: forward and backward.
10. Starting at one, count by fives to six: forward and backward.
11. Starting at two, count by fives to seven: forward and backward.
12. Starting at three, count by fives to eight: forward and backward.
13. Starting at four, count by fives to nine: forward and backward.
14. Starting at five, count by fives to ten: forward and backward.

MORE, LESS, EQUAL
FACILITY EXERCISES #4

Answer each question below:

1. Which is more, one or five?
2. Which is less, two or one?
3. Which is more, two or three?
4. Which is less, four or seven?
5. Which is more, eight or five?
6. Which is less, six or eight?
7. Which is less, eight or three?
8. Which is more, two or seven?
9. Which is more, four or five?
10. Which is less, four or five?
11. Which is more, three or five?
12. Which is more, two or nine?
13. Which is less, eight or four?
14. Which is more, eight or four?
15. Which is less, five or seven?
16. Which is more, nine or three?
17. Which is more, ten or six?
18. Which is less, three or six?
19. Which is less, seven or six?
20. Which is more, ten or eight?
21. Which is less, one or ten?
22. Which is more, three or seven?
23. Which is less, five or three?
24. Which is more, nine or four?

ADDITION FACTS
FACILITY EXERCISES #5

Write your answers in the spaces provided.
You may use fingers if you need to, but **there must be no counting**.

1. $5+3=$ ___

2. $\begin{array}{r} 1 \\ +7 \\ \hline \end{array}$

3. $1+5=$ ___

4. $\begin{array}{r} 6 \\ +4 \\ \hline \end{array}$

5. $2+2=$ ___

6. $3+4=$ ___

7. $2+4=$ ___

8. $\begin{array}{r} 9 \\ +1 \\ \hline \end{array}$

9. $\begin{array}{r} 6 \\ +2 \\ \hline \end{array}$

10. $5+5=$ ___

11. $\begin{array}{r} 4 \\ +6 \\ \hline \end{array}$

12. $\begin{array}{r} 8 \\ +1 \\ \hline \end{array}$

13. $\begin{array}{r} 1 \\ +2 \\ \hline \end{array}$

14. $3+2=$ ___

15. $\begin{array}{r} 3 \\ +7 \\ \hline \end{array}$

16. $7 + 3 =$ ___

17. $\begin{array}{r} 1 \\ \underline{+1} \end{array}$

18. $5 + 4 =$ ___

19. $\begin{array}{r} 3 \\ \underline{+4} \end{array}$

20. $\begin{array}{r} 8 \\ \underline{+2} \end{array}$

21. $6 + 1 =$ ___

22. $\begin{array}{r} 7 \\ \underline{+2} \end{array}$

23. $6 + 3 =$ ___

24. $\begin{array}{r} 5 \\ \underline{+5} \end{array}$

25. $\begin{array}{r} 3 \\ \underline{+3} \end{array}$

26. $4 + 4 =$ ___

27. $2 + 8 =$ ___

28. $4 + 1 =$ ___

29. $\begin{array}{r} 1 \\ \underline{+3} \end{array}$

30. $3 + 6 =$ ___

31. $5 + 2 =$ ___

32. $4 + 2 =$ ___

33. $\begin{array}{r} 5 \\ \underline{+1} \end{array}$

LINKING ADDITION AND SUBTRACTION
FACILITY EXERCISES #6

Write your answers in the spaces provided.
You may use fingers if you need to, but **there must be no counting.**

(1)	(2)	(3)
$3+4=$ ___	$7+2=$ ___	$5+3=$ ___
$7-4=$ ___	$9-2$	$8-5$
$7-3=$ ___	$2+7=$ ___	$3+5=$ ___
$4+3=$ ___	$9-7$	$8-3$

(4)		(5)		(6)	
$8+2$	$10-2$	$6+4$	$4+6$	$3+3$	$4+4$
$10-8$	$2+8$	$10-4$	$10-6$	$6-3=$ ___	
				$8-4=$ ___	

(7)

$6+2=$ ___

$8-6=$ ___

$2+6=$ ___

$8-2=$ ___

(8)

$4+2=$ ___

$6-4=$ ___

$6-2=$ ___

$2+4=$ ___

(9)

$3+6=$ ___

$\begin{array}{r} 9 \\ -6 \\ \hline \end{array}$ $\qquad$ $\begin{array}{r} 9 \\ -3 \\ \hline \end{array}$

$6+3=$ ___

(10)

$4+5=$ ___

$\begin{array}{r} 9 \\ -5 \\ \hline \end{array}$ $\qquad$ $\begin{array}{r} 9 \\ -4 \\ \hline \end{array}$

$5+4=$ ___

(11)

$\begin{array}{r} 5 \\ +5 \\ \hline \end{array}$ $\qquad$ $\begin{array}{r} 2 \\ +2 \\ \hline \end{array}$

$4-2=$ ___

$10-5=$ ___

(12)

$3+2=$ ___

$5-3=$ ___

$\begin{array}{r} 2 \\ +3 \\ \hline \end{array}$

$5-2=$ ___

(13)

$\begin{array}{r} 7 \\ +3 \\ \hline \end{array}$ $\qquad$ $\begin{array}{r} 10 \\ -7 \\ \hline \end{array}$

$3+7=$ ___

$10-3=$ ___

SUBTRACTION FACTS
FACILITY EXERCISES #7

Write your answer to each example in the space provided; then check it. You may use fingers if you need to, but **there must be no counting**.

(1)

6 – 2 = ___

Check

___ + ___ = ___

(2)

Check

$$\begin{array}{r} 5 \\ -1 \\ \hline \end{array}$$

____

(3)

Check

$$\begin{array}{r} 4 \\ -3 \\ \hline \end{array}$$

____

(4)

6 – 4 = ___

Check

___ + ___ = ___

(5)

7 – 4 = ___

Check

___ + ___ = ___

(6)

8 – 7 = ___

Check

___ + ___ = ___

(7)

8 – 6 = ___

Check

___ + ___ = ___

(8)

8 – 5 = ___

Check

___ + ___ = ___

(9)

8 – 4 = ___

Check

___ + ___ = ___

(10)

$8 - 3 =$ ___

Check

___ + ___ = ___

(11)

Check

$$\begin{array}{r} 7 \\ -3 \\ \hline \end{array}$$ ___

(12)

Check

$$\begin{array}{r} 5 \\ -4 \\ \hline \end{array}$$ ___

(13)

Check

$$\begin{array}{r} 9 \\ -3 \\ \hline \end{array}$$ ___

(14)

Check

$$\begin{array}{r} 6 \\ -3 \\ \hline \end{array}$$ ___

(15)

Check

$$\begin{array}{r} 9 \\ -5 \\ \hline \end{array}$$ ___

(16)

Check

$$\begin{array}{r} 2 \\ -1 \\ \hline \end{array}$$ ___

(17)

$5 - 3 =$ ___

Check

___ + ___ = ___

(18)

$4 - 2 =$ ___

Check

___ + ___ = ___

(19)

$9 - 2 =$ ___

Check

___ + ___ = ___

(20)

Check

$$\begin{array}{r} 9 \\ -7 \\ \hline \end{array}$$ ___

(21)

$7 - 5 =$ ___

Check

___ + ___ = ___

Work on each set of exercises below. **There must be no counting.**

(22)	(23)	(24)
9 – 1 = ___	**5 – 4 = ___**	**8 – 3 = ___**
9 – 2 = ___	**6 – 4 = ___**	**7 – 3 = ___**
9 – 3 = ___	**7 – 4 = ___**	**6 – 3 = ___**
9 – 4 = ___	**8 – 4 = ___**	**5 – 3 = ___**
9 – 5 = ___	**9 – 4 = ___**	**4 – 3 = ___**
(25)	**(26)**	**(27)**
8 – 5 = ___	**9 – 7 = ___**	**6 – 2 = ___**
7 – 4 = ___	**8 – 6 = ___**	**7 – 3 = ___**
6 – 3 = ___	**7 – 5 = ___**	**8 – 4 = ___**
5 – 2 = ___	**6 – 4 = ___**	**9 – 5 = ___**
4 – 1 = ___	**5 – 3 = ___**	**10 – 6 = ___**

ADDING AND SUBTRACTING WITH ZERO
FACILITY EXERCISES #8

Write your answers in the spaces provided.

(1)

$6 + 0 =$ ___

$6 - 0 =$ ___

$0 + 6 =$ ___

(2)

$6 - 6 =$ ___

$2 + 0 =$ ___

$8 - 8 =$ ___

(3)

$8 + 0 =$ ___

$2 - 0 =$ ___

$10 - 0 =$ ___

(4)

$$\begin{array}{r} 3 \\ +0 \\ \hline \end{array}$$

(5)

$$\begin{array}{r} 0 \\ +4 \\ \hline \end{array}$$

(6)

$$\begin{array}{r} 9 \\ -9 \\ \hline \end{array}$$

(7)

$$\begin{array}{r} 4 \\ -0 \\ \hline \end{array}$$

(8)

$$\begin{array}{r} 3 \\ -3 \\ \hline \end{array}$$

(9)

$7 - 0 =$ ___

$5 - 0 =$ ___

$0 + 5 =$ ___

(10)

$10 + 0 =$ ___

$1 - 0 =$ ___

$5 + 0 =$ ___

(11)

$10 - 10 =$ ___

$4 - 4 =$ ___

$0 + 8 =$ ___

(12)

$$\begin{array}{r} 7 \\ -7 \\ \hline \end{array}$$

(13)

$$\begin{array}{r} 0 \\ +1 \\ \hline \end{array}$$

(14)

$$\begin{array}{r} 1 \\ -1 \\ \hline \end{array}$$

(15)

$$\begin{array}{r} 0 \\ +0 \\ \hline \end{array}$$

(16)

$$\begin{array}{r} 0 \\ -0 \\ \hline \end{array}$$

SUMS AND DIFFERENCES INVOLVING MORE THAN TWO ADDENDS
FACILITY EXERCISES #9

Write your answers in the spaces provided.
You may use fingers if you need to, but **there must be no counting.**

(1)	(2)
$1+1+1+1=$ ___	$2+2+2=$ ___
$3+3+3=$ ___	$4+1+2=$ ___
$5+0+1=$ ___	$2+5+3=$ ___
(3)	**(4)**
$6-4+3=$ ___	$8-2-2-1+4=$ ___
$3+3-2-2+1=$ ___	$2+2+2-3-3=$ ___
$4-1-2+4-1=$ ___	$5+5-3-2-1=$ ___
(5)	**(6)**
$6+2-2+2-2=$ ___	$4+4-3+2=$ ___
$3+3+3-1=$ ___	$1+2+3+4=$ ___
$10-4-3-3=$ ___	$7-0+3-3+0=$ ___
(7)	**(8)**
$0+0+0+0=$ ___	$10-3-3-3=$ ___
$0-0-0-0=$ ___	$10-5-5=$ ___
$6-2-2-2=$ ___	$0+9-4-4=$ ___

(9)	(10)
$0-0+0-0=$ ___	$7-7+7-7=$ ___
$9-2+2=$ ___	$10-9+8-7=$ ___
$10+6-6=$ ___	$9-2-3+5=$ ___

Draw a line under the correct answer in each question below (#11, #12 and #13).

(11)

Which is larger: $2+5$ or $1+2+1+2$?

Which is larger: $3+2+3$ or $5+5$?

Which is larger: $1+1+1+1$ or 7?

(12)

Which is less: $9-3-3-3$ or 2?

Which is less: $5-0-1+3$ or $2+2+2+2$?

Which is larger: 9 or $3+6-5-1$?

(13)

Which is larger: $2+3+1$ or $10-1-2$?

Which is less: 8 or $3+3+4-1$?

Which is less: $7+0-1$ or $5+1$?

THE LOWER ADDITION AND SUBTRACTION FACTS
FACILITY EXERCISES #10

Answer the problems below. There must be **no use of fingers or any counting.**

1. $6 + 4 =$ __
2. $10 - 5$
3. $8 + 2 =$ __
4. $3 + 7$
5. $5 + 5$
6. $10 - 2$
7. $10 - 6 =$ __
8. $10 - 7$
9. $9 + 1$
10. $4 + 6 =$ __
11. $10 - 5$
12. $2 + 8 =$ __
13. $7 + 3 =$ __
14. $10 - 4$
15. $10 - 3$
16. $3 + 7 =$ __

THE LOWER ADDITION AND SUBTRACTION FACTS
FACILITY EXERCISES #11

Answer the problems below. There must be **no use of fingers or any counting.**

1. $8 + 2$
2. $7 + 3 =$ __
3. $10 - 9$
4. $7 + 3$
5. $10 - 6 =$ __
6. $10 - 8$
7. $1 + 9 =$ __
8. $6 + 4$
9. $10 - 1$
10. $10 - 4$
11. $5 + 5$
12. $3 + 7 =$ __
13. $10 - 5 =$ __
14. $10 - 7 =$ __
15. $2 + 8$
16. $10 - 3$

THE LOWER ADDITION AND SUBTRACTION FACTS
FACILITY EXERCISES #12

Answer the problems below. There must be **no use of fingers or any counting.**

1. $\begin{array}{r} 10 \\ -\ 1 \\ \hline \end{array}$
2. $3+7=$ __
3. $\begin{array}{r} 6 \\ +4 \\ \hline \end{array}$
4. $\begin{array}{r} 2 \\ +8 \\ \hline \end{array}$
5. $10-5=$ __
6. $10-2=$ __
7. $\begin{array}{r} 10 \\ -\ 3 \\ \hline \end{array}$
8. $\begin{array}{r} 1 \\ +9 \\ \hline \end{array}$
9. $8+2=$ __
10. $\begin{array}{r} 10 \\ -\ 7 \\ \hline \end{array}$
11. $5+5=$ __
12. $\begin{array}{r} 7 \\ +3 \\ \hline \end{array}$
13. $\begin{array}{r} 10 \\ -\ 9 \\ \hline \end{array}$
14. $10-4=$ __
15. $9+1=$ __
16. $4+6=$ __

THE LOWER ADDITION AND SUBTRACTION FACTS
FACILITY EXERCISES #13

Answer the problems below. There must be **no use of fingers or any counting.**

1. $10-8=$ __
2. $1+9=$ __
3. $\begin{array}{r} 8 \\ +2 \\ \hline \end{array}$
4. $\begin{array}{r} 10 \\ -\ 6 \\ \hline \end{array}$
5. $10-3=$ __
6. $\begin{array}{r} 3 \\ +7 \\ \hline \end{array}$
7. $\begin{array}{r} 10 \\ -\ 1 \\ \hline \end{array}$
8. $10-8=$ __
9. $4+6=$ __
10. $\begin{array}{r} 9 \\ +1 \\ \hline \end{array}$
11. $10-9=$ __
12. $\begin{array}{r} 5 \\ +5 \\ \hline \end{array}$
13. $\begin{array}{r} 10 \\ -\ 7 \\ \hline \end{array}$
14. $2+8=$ __
15. $10-5=$ __
16. $\begin{array}{r} 10 \\ -\ 3 \\ \hline \end{array}$

THE LOWER ADDITION AND SUBTRACTION FACTS
FACILITY EXERCISES #14

Answer the problems below. There must be **no use of fingers or any counting.**

1. $5+5=$__
2. $\begin{array}{r} 10 \\ -\ 4 \\ \hline \end{array}$
3. $10-7=$__
4. $\begin{array}{r} 10 \\ -\ 9 \\ \hline \end{array}$
5. $7+3=$__
6. $\begin{array}{r} 1 \\ +9 \\ \hline \end{array}$
7. $\begin{array}{r} 10 \\ -\ 6 \\ \hline \end{array}$
8. $8+2=$__
9. $10-4=$__
10. $6+4=$__
11. $\begin{array}{r} 3 \\ +7 \\ \hline \end{array}$
12. $\begin{array}{r} 10 \\ -\ 5 \\ \hline \end{array}$
13. $\begin{array}{r} 8 \\ +2 \\ \hline \end{array}$
14. $\begin{array}{r} 10 \\ -\ 3 \\ \hline \end{array}$
15. $9+1=$__
16. $\begin{array}{r} 4 \\ +6 \\ \hline \end{array}$

THE LOWER ADDITION AND SUBTRACTION FACTS
FACILITY EXERCISES #15

Answer the problems below. There must be **no use of fingers or any counting.**

1. $\begin{array}{r} 8 \\ +2 \\ \hline \end{array}$
2. $6+4=$__
3. $10-1=$__
4. $2+8=$__
5. $\begin{array}{r} 7 \\ +3 \\ \hline \end{array}$
6. $\begin{array}{r} 10 \\ -\ 6 \\ \hline \end{array}$
7. $9+1=$__
8. $10-2=$__
9. $10-4=$__
10. $\begin{array}{r} 10 \\ -\ 5 \\ \hline \end{array}$
11. $\begin{array}{r} 4 \\ +6 \\ \hline \end{array}$
12. $\begin{array}{r} 10 \\ -\ 8 \\ \hline \end{array}$
13. $\begin{array}{r} 9 \\ +1 \\ \hline \end{array}$
14. $5+5=$__
15. $\begin{array}{r} 10 \\ -\ 3 \\ \hline \end{array}$
16. $10-6=$__

THE LOWER ADDITION AND SUBTRACTION FACTS
FACILITY EXERCISES #16

Answer the problems below. There must be **no use of fingers or any counting.**

1. $9+1=$__
2. $\begin{array}{r} 6 \\ +4 \\ \hline \end{array}$
3. $10-7=$__
4. $\begin{array}{r} 10 \\ -\ 9 \\ \hline \end{array}$
5. $7+3=$__
6. $\begin{array}{r} 2 \\ +8 \\ \hline \end{array}$
7. $4+6=$__
8. $\begin{array}{r} 10 \\ -\ 6 \\ \hline \end{array}$
9. $10-1=$__
10. $\begin{array}{r} 5 \\ +5 \\ \hline \end{array}$
11. $\begin{array}{r} 3 \\ +7 \\ \hline \end{array}$
12. $\begin{array}{r} 10 \\ -\ 8 \\ \hline \end{array}$
13. $\begin{array}{r} 10 \\ -\ 4 \\ \hline \end{array}$
14. $10-6=$__
15. $\begin{array}{r} 9 \\ +1 \\ \hline \end{array}$
16. $5+5=$__

THE LOWER ADDITION AND SUBTRACTION FACTS
FACILITY EXERCISES #17

Answer the problems below. There must be **no use of fingers or any counting.**

1. $7 + 3 =$__
2. $10 - 1 =$__
3. $5 + 5 =$__
4. $9 + 1 =$__
5. $\begin{array}{r} 10 \\ -5 \\ \hline \end{array}$
6. $10 - 6 =$__
7. $\begin{array}{r} 8 \\ +2 \\ \hline \end{array}$
8. $3 + 7 =$__
9. $6 + 4 =$__
10. $\begin{array}{r} 10 \\ -\ 9 \\ \hline \end{array}$
11. $\begin{array}{r} 10 \\ -\ 3 \\ \hline \end{array}$
12. $10 - 3 =$__
13. $\begin{array}{r} 9 \\ +1 \\ \hline \end{array}$
14. $\begin{array}{r} 4 \\ +6 \\ \hline \end{array}$
15. $10 - 7 =$__
16. $\begin{array}{r} 2 \\ +8 \\ \hline \end{array}$

THE LOWER ADDITION AND SUBTRACTION FACTS
FACILITY EXERCISES #18

Answer the problems below. There must be **no use of fingers or any counting.**

1. $5 + 4 =$ __

2. $\begin{array}{r} 9 \\ -\,8 \\ \hline \end{array}$

3. $10 - 7 =$ __

4. $8 + 1 =$ __

5. $\begin{array}{r} 5 \\ +\,5 \\ \hline \end{array}$

6. $\begin{array}{r} 7 \\ +\,2 \\ \hline \end{array}$

7. $9 - 1 =$ __

8. $\begin{array}{r} 9 \\ -\,3 \\ \hline \end{array}$

9. $\begin{array}{r} 2 \\ +\,8 \\ \hline \end{array}$

10. $6 + 3 =$ __

11. $\begin{array}{r} 10 \\ -\,4 \\ \hline \end{array}$

12. $9 - 5 =$ __

13. $9 - 7 =$ __

14. $3 + 7 =$ __

15. $\begin{array}{r} 3 \\ +\,6 \\ \hline \end{array}$

16. $\begin{array}{r} 9 \\ -\,2 \\ \hline \end{array}$

17. $9 - 6 =$ __

18. $\begin{array}{r} 9 \\ -\,4 \\ \hline \end{array}$

19. $4 + 6 =$ __

20. $2 + 7 =$ __

THE LOWER ADDITION AND SUBTRACTION FACTS
FACILITY EXERCISES #19

Answer the problems below. There must be **no use of fingers or any counting**.

1. $\begin{array}{r} 7 \\ \underline{+2} \end{array}$

2. $9 - 3 = \_\_$

3. $8 + 2 = \_\_$

4. $\begin{array}{r} 10 \\ \underline{-6} \end{array}$

5. $1 + 8 = \_\_$

6. $9 - 7 = \_\_$

7. $\begin{array}{r} 4 \\ \underline{+5} \end{array}$

8. $\begin{array}{r} 3 \\ \underline{+7} \end{array}$

9. $10 - 2 = \_\_$

10. $\begin{array}{r} 9 \\ \underline{-4} \end{array}$

11. $\begin{array}{r} 10 \\ \underline{-9} \end{array}$

12. $\begin{array}{r} 9 \\ \underline{-6} \end{array}$

13. $\begin{array}{r} 6 \\ \underline{+4} \end{array}$

14. $\begin{array}{r} 9 \\ \underline{-7} \end{array}$

15. $9 - 5 = \_\_$

16. $8 + 1 = \_\_$

17. $6 + 3 = \_\_$

18. $5 + 4 = \_\_$

19. $\begin{array}{r} 10 \\ \underline{-1} \end{array}$

20. $10 - 8 = \_\_$

THE LOWER ADDITION AND SUBTRACTION FACTS
FACILITY EXERCISES #20

Answer the problems below. There must be **no use of fingers or any counting.**

1. $3 + 6 =$ __

2. $\begin{array}{r} 9 \\ -6 \\ \hline \end{array}$

3. $\begin{array}{r} 10 \\ -4 \\ \hline \end{array}$

4. $9 - 2 =$ __

5. $8 + 2 =$ __

6. $\begin{array}{r} 4 \\ +5 \\ \hline \end{array}$

7. $9 - 8 =$ __

8. $10 - 7 =$ __

9. $\begin{array}{r} 9 \\ -7 \\ \hline \end{array}$

10. $\begin{array}{r} 6 \\ +3 \\ \hline \end{array}$

11. $1 + 8 =$ __

12. $\begin{array}{r} 10 \\ -2 \\ \hline \end{array}$

13. $7 + 2 =$ __

14. $\begin{array}{r} 9 \\ -3 \\ \hline \end{array}$

15. $9 - 7 =$ __

16. $\begin{array}{r} 9 \\ -5 \\ \hline \end{array}$

17. $\begin{array}{r} 5 \\ +5 \\ \hline \end{array}$

18. $9 - 4 =$ __

19. $10 - 6 =$ __

20. $\begin{array}{r} 10 \\ -1 \\ \hline \end{array}$

THE LOWER ADDITION AND SUBTRACTION FACTS
FACILITY EXERCISES #21

Answer the problems below. There must be **no use of fingers or any counting**.

1. $1 + 8 =$ ___
2. $9 - 5$
3. $10 - 6$
4. $9 - 7 =$ ___
5. $9 - 4$
6. $7 + 2 =$ ___
7. $10 - 8 =$ ___
8. $6 + 3$
9. $9 - 2$
10. $9 + 1 =$ ___
11. $10 - 3 =$ ___
12. $2 + 8 =$ ___
13. $7 + 2$
14. $6 + 4$
15. $9 - 2 =$ ___
16. $8 + 1$
17. $5 + 4$
18. $9 - 3 =$ ___
19. $10 - 1$
20. $3 + 7 =$ ___

THE LOWER ADDITION AND SUBTRACTION FACTS
FACILITY EXERCISES #22

Answer the problems below. There must be **no use of fingers or any counting.**

1. $\begin{array}{r} 10 \\ -\ 2 \\ \hline \end{array}$

2. $9 - 4 =$ __

3. $\begin{array}{r} 6 \\ +\ 3 \\ \hline \end{array}$

4. $5 + 4 =$ __

5. $10 - 7 =$ __

6. $\begin{array}{r} 2 \\ +\ 7 \\ \hline \end{array}$

7. $\begin{array}{r} 9 \\ -\ 5 \\ \hline \end{array}$

8. $\begin{array}{r} 9 \\ -\ 3 \\ \hline \end{array}$

9. $\begin{array}{r} 10 \\ -\ 6 \\ \hline \end{array}$

10. $9 - 5 =$ __

11. $10 - 8 =$ __

12. $\begin{array}{r} 9 \\ -\ 7 \\ \hline \end{array}$

13. $\begin{array}{r} 9 \\ -\ 6 \\ \hline \end{array}$

14. $2 + 7 =$ __

15. $\begin{array}{r} 4 \\ +\ 6 \\ \hline \end{array}$

16. $\begin{array}{r} 9 \\ -\ 8 \\ \hline \end{array}$

17. $10 - 5 =$ __

18. $6 + 3 =$ __

19. $\begin{array}{r} 10 \\ -\ 1 \\ \hline \end{array}$

20. $9 - 6 =$ __

THE LOWER ADDITION AND SUBTRACTION FACTS
FACILITY EXERCISES #23

Answer the problems below. There must be **no use of fingers or any counting.**

1. $\begin{array}{r} 7 \\ +2 \\ \hline \end{array}$

2. $\begin{array}{r} 10 \\ -4 \\ \hline \end{array}$

3. $8 + 1 =$ __

4. $6 + 3 =$ __

5. $2 + 8 =$ __

6. $5 + 4 =$ __

7. $\begin{array}{r} 9 \\ -2 \\ \hline \end{array}$

8. $\begin{array}{r} 9 \\ -1 \\ \hline \end{array}$

9. $\begin{array}{r} 6 \\ +4 \\ \hline \end{array}$

10. $\begin{array}{r} 9 \\ -7 \\ \hline \end{array}$

11. $10 - 5 =$ __

12. $\begin{array}{r} 3 \\ +6 \\ \hline \end{array}$

13. $2 + 7 =$ __

14. $9 - 8 =$ __

15. $\begin{array}{r} 9 \\ -3 \\ \hline \end{array}$

16. $\begin{array}{r} 4 \\ +5 \\ \hline \end{array}$

17. $\begin{array}{r} 10 \\ -7 \\ \hline \end{array}$

18. $9 - 4 =$ __

19. $10 - 6 =$ __

20. $\begin{array}{r} 7 \\ +3 \\ \hline \end{array}$

THE LOWER ADDITION AND SUBTRACTION FACTS
FACILITY EXERCISES #24

Answer the problems below. There must be **no use of fingers or any counting.**

1. $2 + 7 =$ __

2. $4 + 6 =$ __

3. $9 - 4$

4. $9 - 6$

5. $9 - 2$

6. $3 + 6$

7. $3 + 7 =$ __

8. $9 - 7 =$ __

9. $9 - 5 =$ __

10. $10 - 4$

11. $6 + 3 =$ __

12. $2 + 8$

13. $9 - 3$

14. $9 - 1 =$ __

15. $7 + 2$

16. $5 + 5$

17. $8 + 1 =$ __

18. $5 + 4$

19. $9 - 8$

20. $10 - 7 =$ __

THE LOWER ADDITION AND SUBTRACTION FACTS
FACILITY EXERCISES #25

Answer the problems below. There must be **no use of fingers or any counting.**

1. $\begin{array}{r} 9 \\ -8 \\ \hline \end{array}$

2. $8+2=$ __

3. $9-2=$ __

4. $9-7=$ __

5. $\begin{array}{r} 10 \\ -2 \\ \hline \end{array}$

6. $5+4=$ __

7. $3+7=$ __

8. $\begin{array}{r} 9 \\ -3 \\ \hline \end{array}$

9. $\begin{array}{r} 9 \\ -7 \\ \hline \end{array}$

10. $\begin{array}{r} 9 \\ -4 \\ \hline \end{array}$

11. $1+8=$ __

12. $9-5=$ __

13. $2+7=$ __

14. $9-1=$ __

15. $\begin{array}{r} 3 \\ +6 \\ \hline \end{array}$

16. $8+1=$ __

17. $9-6=$ __

18. $9-4=$ __

19. $10-6=$ __

20. $\begin{array}{r} 4 \\ +6 \\ \hline \end{array}$

THE LOWER ADDITION AND SUBTRACTION FACTS
FACILITY EXERCISES #26

Answer the problems below. There must be **no use of fingers or any counting.**

1. $\begin{array}{r} 6 \\ +2 \\ \hline \end{array}$

2. $8 - 5 =$ ___

3. $5 + 3 =$ ___

4. $\begin{array}{r} 9 \\ -6 \\ \hline \end{array}$

5. $\begin{array}{r} 8 \\ -4 \\ \hline \end{array}$

6. $\begin{array}{r} 6 \\ +4 \\ \hline \end{array}$

7. $7 + 1 =$ ___

8. $8 - 3 =$ ___

9. $8 - 6 =$ ___

10. $\begin{array}{r} 9 \\ -2 \\ \hline \end{array}$

11. $\begin{array}{r} 3 \\ +7 \\ \hline \end{array}$

12. $\begin{array}{r} 8 \\ -7 \\ \hline \end{array}$

13. $10 - 2 =$ ___

14. $3 + 5 =$ ___

15. $8 - 1 =$ ___

16. $\begin{array}{r} 4 \\ +4 \\ \hline \end{array}$

17. $\begin{array}{r} 9 \\ -5 \\ \hline \end{array}$

18. $\begin{array}{r} 8 \\ -2 \\ \hline \end{array}$

19. $10 - 7 =$ ___

20. $2 + 6 =$ ___

THE LOWER ADDITION AND SUBTRACTION FACTS
FACILITY EXERCISES #27

Answer the problems below. There must be **no use of fingers or any counting**.

1. $4 + 4 =$ __

2. $\begin{array}{r} 8 \\ -\,2 \\ \hline \end{array}$

3. $\begin{array}{r} 7 \\ +\,1 \\ \hline \end{array}$

4. $8 - 4 =$ __

5. $6 + 2 =$ __

6. $\begin{array}{r} 8 \\ +\,2 \\ \hline \end{array}$

7. $8 - 5 =$ __

8. $\begin{array}{r} 7 \\ +\,2 \\ \hline \end{array}$

9. $\begin{array}{r} 8 \\ -\,7 \\ \hline \end{array}$

10. $6 + 3 =$ __

11. $8 - 6 =$ __

12. $\begin{array}{r} 10 \\ -\,7 \\ \hline \end{array}$

13. $8 - 2 =$ __

14. $\begin{array}{r} 8 \\ -\,5 \\ \hline \end{array}$

15. $2 + 6 =$ __

16. $3 + 5 =$ __

17. $\begin{array}{r} 9 \\ -\,7 \\ \hline \end{array}$

18. $\begin{array}{r} 4 \\ +\,5 \\ \hline \end{array}$

19. $8 - 3 =$ __

20. $\begin{array}{r} 10 \\ -\,4 \\ \hline \end{array}$

THE LOWER ADDITION AND SUBTRACTION FACTS
FACILITY EXERCISES #28

Answer the problems below. There must be **no use of fingers or any counting.**

1. $8 - 5 =$ __

2. $\begin{array}{r} 2 \\ +\,6 \\ \hline \end{array}$

3. $8 + 2 =$ __

4. $\begin{array}{r} 8 \\ -\,6 \\ \hline \end{array}$

5. $3 + 5 =$ __

6. $\begin{array}{r} 10 \\ -\,3 \\ \hline \end{array}$

7. $\begin{array}{r} 8 \\ -\,3 \\ \hline \end{array}$

8. $4 + 4 =$ __

9. $8 - 6 =$ __

10. $\begin{array}{r} 9 \\ -\,5 \\ \hline \end{array}$

11. $6 + 2 =$ __

12. $\begin{array}{r} 8 \\ -\,5 \\ \hline \end{array}$

13. $\begin{array}{r} 8 \\ -\,2 \\ \hline \end{array}$

14. $5 + 4 =$ __

15. $8 - 7 =$ __

16. $3 + 6 =$ __

17. $\begin{array}{r} 10 \\ -\,7 \\ \hline \end{array}$

18. $7 + 1 =$ __

19. $\begin{array}{r} 9 \\ -\,2 \\ \hline \end{array}$

20. $\begin{array}{r} 2 \\ +\,8 \\ \hline \end{array}$

THE LOWER ADDITION AND SUBTRACTION FACTS
FACILITY EXERCISES #29

Answer the problems below. There must be **no use of fingers or any counting.**

1. $3 + 5 =$ __

2. $\begin{array}{r} 2 \\ +6 \\ \hline \end{array}$

3. $8 - 5 =$ __

4. $8 - 2 =$ __

5. $\begin{array}{r} 10 \\ -7 \\ \hline \end{array}$

6. $\begin{array}{r} 8 \\ -6 \\ \hline \end{array}$

7. $\begin{array}{r} 6 \\ +3 \\ \hline \end{array}$

8. $8 - 7 =$ __

9. $\begin{array}{r} 7 \\ +2 \\ \hline \end{array}$

10. $8 - 1 =$ __

11. $\begin{array}{r} 8 \\ +2 \\ \hline \end{array}$

12. $6 + 2 =$ __

13. $8 - 4 =$ __

14. $\begin{array}{r} 7 \\ +1 \\ \hline \end{array}$

15. $\begin{array}{r} 8 \\ -2 \\ \hline \end{array}$

16. $4 + 4 =$ __

17. $\begin{array}{r} 10 \\ -4 \\ \hline \end{array}$

18. $\begin{array}{r} 8 \\ -3 \\ \hline \end{array}$

19. $\begin{array}{r} 4 \\ +5 \\ \hline \end{array}$

20. $9 - 7 =$ __

THE LOWER ADDITION AND SUBTRACTION FACTS
FACILITY EXERCISES #30

Answer the problems below. There must be **no use of fingers or any counting.**

1. $\begin{array}{r} 8 \\ -1 \\ \hline \end{array}$

2. $4 + 4 = \_\_$

3. $\begin{array}{r} 9 \\ -3 \\ \hline \end{array}$

4. $\begin{array}{r} 5 \\ +3 \\ \hline \end{array}$

5. $\begin{array}{r} 10 \\ -2 \\ \hline \end{array}$

6. $\begin{array}{r} 2 \\ +6 \\ \hline \end{array}$

7. $\begin{array}{r} 8 \\ -7 \\ \hline \end{array}$

8. $8 - 4 = \_\_$

9. $\begin{array}{r} 5 \\ +5 \\ \hline \end{array}$

10. $3 + 5 = \_\_$

11. $\begin{array}{r} 8 \\ -6 \\ \hline \end{array}$

12. $8 - 1 = \_\_$

13. $6 + 3 = \_\_$

14. $\begin{array}{r} 8 \\ -2 \\ \hline \end{array}$

15. $10 - 3 = \_\_$

16. $8 - 5 = \_\_$

17. $7 + 1 = \_\_$

18. $\begin{array}{r} 10 \\ -6 \\ \hline \end{array}$

19. $8 - 3 = \_\_$

20. $\begin{array}{r} 5 \\ +4 \\ \hline \end{array}$

THE LOWER ADDITION AND SUBTRACTION FACTS
FACILITY EXERCISES #31

Answer the problems below. There must be **no use of fingers or any counting**.

1. $\begin{array}{r} 8 \\ -\ 4 \\ \hline \end{array}$

2. $\begin{array}{r} 7 \\ +\ 1 \\ \hline \end{array}$

3. $8 - 3 =$ __

4. $\begin{array}{r} 4 \\ +\ 4 \\ \hline \end{array}$

5. $7 + 2 =$ __

6. $8 - 2 =$ __

7. $\begin{array}{r} 8 \\ +\ 2 \\ \hline \end{array}$

8. $\begin{array}{r} 6 \\ +\ 2 \\ \hline \end{array}$

9. $\begin{array}{r} 10 \\ -\ 7 \\ \hline \end{array}$

10. $8 - 6 =$ __

11. $6 + 3 =$ __

12. $\begin{array}{r} 8 \\ -\ 7 \\ \hline \end{array}$

13. $8 - 5 =$ __

14. $5 + 3 =$ __

15. $2 + 6 =$ __

16. $\begin{array}{r} 8 \\ -\ 5 \\ \hline \end{array}$

17. $\begin{array}{r} 10 \\ -\ 4 \\ \hline \end{array}$

18. $\begin{array}{r} 8 \\ -\ 3 \\ \hline \end{array}$

19. $\begin{array}{r} 4 \\ +\ 5 \\ \hline \end{array}$

20. $\begin{array}{r} 9 \\ -\ 7 \\ \hline \end{array}$

THE LOWER ADDITION AND SUBTRACTION FACTS
FACILITY EXERCISES #32

Answer the problems below. There must be **no use of fingers or any counting.**

1. $8 - 7 = \_\_$
2. $\begin{array}{r} 6 \\ +2 \\ \hline \end{array}$
3. $\begin{array}{r} 8 \\ -1 \\ \hline \end{array}$
4. $10 - 8 = \_\_$
5. $\begin{array}{r} 8 \\ -5 \\ \hline \end{array}$
6. $\begin{array}{r} 4 \\ +5 \\ \hline \end{array}$
7. $5 + 3 = \_\_$
8. $9 - 6 = \_\_$
9. $10 - 3 = \_\_$
10. $\begin{array}{r} 7 \\ +1 \\ \hline \end{array}$
11. $4 + 4 = \_\_$
12. $2 + 6 = \_\_$
13. $\begin{array}{r} 3 \\ +5 \\ \hline \end{array}$
14. $\begin{array}{r} 8 \\ -6 \\ \hline \end{array}$
15. $8 - 1 = \_\_$
16. $\begin{array}{r} 8 \\ -3 \\ \hline \end{array}$
17. $\begin{array}{r} 8 \\ -4 \\ \hline \end{array}$
18. $2 + 7 = \_\_$
19. $8 - 2 = \_\_$
20. $\begin{array}{r} 6 \\ +3 \\ \hline \end{array}$

THE LOWER ADDITION AND SUBTRACTION FACTS
FACILITY EXERCISES #33

Answer the problems below. There must be **no use of fingers or any counting**.

1. $\begin{array}{r} 8 \\ -5 \\ \hline \end{array}$

2. $2 + 6 = \_\_$

3. $\begin{array}{r} 9 \\ -7 \\ \hline \end{array}$

4. $\begin{array}{r} 8 \\ -3 \\ \hline \end{array}$

5. $6 + 2 = \_\_$

6. $\begin{array}{r} 8 \\ +2 \\ \hline \end{array}$

7. $\begin{array}{r} 8 \\ -4 \\ \hline \end{array}$

8. $8 - 7 = \_\_$

9. $4 + 4 = \_\_$

10. $1 + 7 = \_\_$

11. $\begin{array}{r} 3 \\ +5 \\ \hline \end{array}$

12. $\begin{array}{r} 8 \\ -2 \\ \hline \end{array}$

13. $\begin{array}{r} 10 \\ -9 \\ \hline \end{array}$

14. $8 - 6 = \_\_$

15. $7 + 2 = \_\_$

16. $9 - 8 = \_\_$

17. $8 - 2 = \_\_$

18. $\begin{array}{r} 3 \\ +7 \\ \hline \end{array}$

19. $\begin{array}{r} 9 \\ -6 \\ \hline \end{array}$

20. $8 - 5 = \_\_$

THE LOWER ADDITION AND SUBTRACTION FACTS
FACILITY EXERCISES #34

Answer the problems below. There must be **no use of fingers or any counting.**

1. $5 + 2 =$ __
2. $\begin{array}{r} 7 \\ -\ 3 \\ \hline \end{array}$
3. $\begin{array}{r} 7 \\ +\ 3 \\ \hline \end{array}$
4. $6 + 1 =$ __
5. $\begin{array}{r} 4 \\ +\ 3 \\ \hline \end{array}$
6. $7 + 2 =$ __
7. $7 - 2 =$ __
8. $\begin{array}{r} 2 \\ +\ 5 \\ \hline \end{array}$
9. $6 + 3 =$ __
10. $\begin{array}{r} 3 \\ +\ 5 \\ \hline \end{array}$
11. $7 - 5 =$ __
12. $\begin{array}{r} 9 \\ -\ 5 \\ \hline \end{array}$
13. $10 - 4 =$ __
14. $\begin{array}{r} 7 \\ -\ 6 \\ \hline \end{array}$
15. $1 + 6 =$ __
16. $\begin{array}{r} 7 \\ -\ 1 \\ \hline \end{array}$
17. $3 + 4 =$ __
18. $8 - 6 =$ __
19. $\begin{array}{r} 7 \\ -\ 4 \\ \hline \end{array}$
20. $\begin{array}{r} 6 \\ +\ 2 \\ \hline \end{array}$

THE LOWER ADDITION AND SUBTRACTION FACTS
FACILITY EXERCISES #35

Answer the problems below. There must be **no use of fingers or any counting.**

1. $\begin{array}{r} 4 \\ +3 \\ \hline \end{array}$

2. $7 - 2 =$ ___

3. $\begin{array}{r} 6 \\ +1 \\ \hline \end{array}$

4. $7 - 4 =$ ___

5. $\begin{array}{r} 7 \\ -1 \\ \hline \end{array}$

6. $\begin{array}{r} 3 \\ +5 \\ \hline \end{array}$

7. $\begin{array}{r} 9 \\ -7 \\ \hline \end{array}$

8. $5 - 2 =$ ___

9. $7 - 6 =$ ___

10. $\begin{array}{r} 8 \\ +2 \\ \hline \end{array}$

11. $\begin{array}{r} 7 \\ -5 \\ \hline \end{array}$

12. $7 - 3 =$ ___

13. $\begin{array}{r} 7 \\ -2 \\ \hline \end{array}$

14. $8 - 5 =$ ___

15. $3 + 4 =$ ___

16. $\begin{array}{r} 7 \\ -4 \\ \hline \end{array}$

17. $5 + 4 =$ ___

18. $\begin{array}{r} 5 \\ +2 \\ \hline \end{array}$

19. $\begin{array}{r} 7 \\ -3 \\ \hline \end{array}$

20. $\begin{array}{r} 10 \\ -4 \\ \hline \end{array}$

THE LOWER ADDITION AND SUBTRACTION FACTS
FACILITY EXERCISES #36

Answer the problems below. There must be **no use of fingers or any counting.**

1. $\begin{array}{r} 6 \\ +1 \\ \hline \end{array}$

2. $4 + 3 =$ __

3. $7 - 5 =$ __

4. $\begin{array}{r} 7 \\ -1 \\ \hline \end{array}$

5. $\begin{array}{r} 3 \\ +5 \\ \hline \end{array}$

6. $\begin{array}{r} 7 \\ -6 \\ \hline \end{array}$

7. $9 - 7 =$ __

8. $\begin{array}{r} 7 \\ -4 \\ \hline \end{array}$

9. $2 + 5 =$ __

10. $\begin{array}{r} 4 \\ +4 \\ \hline \end{array}$

11. $7 - 6 =$ __

12. $\begin{array}{r} 7 \\ -2 \\ \hline \end{array}$

13. $\begin{array}{r} 9 \\ -2 \\ \hline \end{array}$

14. $7 - 3 =$ __

15. $\begin{array}{r} 8 \\ -6 \\ \hline \end{array}$

16. $3 + 4 =$ __

17. $7 - 2 =$ __

18. $\begin{array}{r} 3 \\ +6 \\ \hline \end{array}$

19. $\begin{array}{r} 5 \\ +4 \\ \hline \end{array}$

20. $8 - 5 =$ __

THE LOWER ADDITION AND SUBTRACTION FACTS
FACILITY EXERCISES #37

Answer the problems below. There must be **no use of fingers or any counting.**

1. $2 + 5 =$ __
2. $\begin{array}{r} 4 \\ \underline{+3} \end{array}$
3. $7 - 1 =$ __
4. $10 - 2 =$ __
5. $\begin{array}{r} 9 \\ \underline{-8} \end{array}$
6. $6 + 1 =$ __
7. $7 - 4 =$ __
8. $\begin{array}{r} 5 \\ \underline{+2} \end{array}$
9. $\begin{array}{r} 8 \\ \underline{-5} \end{array}$
10. $9 - 7 =$ __
11. $\begin{array}{r} 7 \\ \underline{-6} \end{array}$
12. $\begin{array}{r} 1 \\ \underline{+7} \end{array}$
13. $\begin{array}{r} 7 \\ \underline{-3} \end{array}$
14. $2 + 6 =$ __
15. $5 + 2 =$ __
16. $7 - 3 =$ __
17. $7 - 5 =$ __
18. $\begin{array}{r} 10 \\ \underline{-6} \end{array}$
19. $\begin{array}{r} 4 \\ \underline{+4} \end{array}$
20. $9 - 3 =$ __

THE LOWER ADDITION AND SUBTRACTION FACTS
FACILITY EXERCISES #38

Answer the problems below. There must be **no use of fingers or any counting**.

1. $8 - 3 =$ __

2. $\begin{array}{r} 7 \\ -\ 5 \\ \hline \end{array}$

3. $\begin{array}{r} 3 \\ +\ 4 \\ \hline \end{array}$

4. $\begin{array}{r} 9 \\ -\ 3 \\ \hline \end{array}$

5. $7 + 3 =$ __

6. $\begin{array}{r} 10 \\ -\ 2 \\ \hline \end{array}$

7. $\begin{array}{r} 9 \\ -\ 5 \\ \hline \end{array}$

8. $1 + 6 =$ __

9. $\begin{array}{r} 8 \\ +\ 1 \\ \hline \end{array}$

10. $9 - 6 =$ __

11. $4 + 3 =$ __

12. $7 - 5 =$ __

13. $\begin{array}{r} 5 \\ +\ 2 \\ \hline \end{array}$

14. $7 - 4 =$ __

15. $\begin{array}{r} 7 \\ -1 \\ \hline \end{array}$

16. $2 + 5 =$ __

17. $7 - 6 =$ __

18. $4 + 5 =$ __

19. $5 + 5 =$ __

20. $2 + 7 =$ __

THE LOWER ADDITION AND SUBTRACTION FACTS
FACILITY EXERCISES #39

Answer the problems below. There must be **no use of fingers or any counting**.

1. $9 - 5 = \_\_$

2. $\begin{array}{r} 7 \\ -\ 2 \\ \hline \end{array}$

3. $1 + 9 = \_\_$

4. $\begin{array}{r} 7 \\ -\ 6 \\ \hline \end{array}$

5. $\begin{array}{r} 3 \\ +\ 5 \\ \hline \end{array}$

6. $\begin{array}{r} 10 \\ -\ 4 \\ \hline \end{array}$

7. $7 - 3 = \_\_$

8. $7 + 2 = \_\_$

9. $7 - 1 = \_\_$

10. $\begin{array}{r} 4 \\ +\ 4 \\ \hline \end{array}$

11. $7 - 5 = \_\_$

12. $8 + 2 = \_\_$

13. $5 + 2 = \_\_$

14. $7 - 2 = \_\_$

15. $\begin{array}{r} 7 \\ -\ 4 \\ \hline \end{array}$

16. $\begin{array}{r} 1 \\ +\ 6 \\ \hline \end{array}$

17. $3 + 4 = \_\_$

18. $\begin{array}{r} 8 \\ -\ 1 \\ \hline \end{array}$

19. $\begin{array}{r} 2 \\ +\ 5 \\ \hline \end{array}$

20. $2 + 5 = \_\_$

THE LOWER ADDITION AND SUBTRACTION FACTS
FACILITY EXERCISES #40

Answer the problems below. There must be **no use of fingers or any counting.**

1. $\begin{array}{r} 8 \\ -2 \\ \hline \end{array}$

2. $4 + 3 =$ __

3. $7 - 5 =$ __

4. $\begin{array}{r} 3 \\ +7 \\ \hline \end{array}$

5. $\begin{array}{r} 4 \\ +6 \\ \hline \end{array}$

6. $8 + 2 =$ __

7. $\begin{array}{r} 5 \\ +4 \\ \hline \end{array}$

8. $2 + 5 =$ __

9. $7 - 4 =$ __

10. $7 - 2 =$ __

11. $7 - 6 =$ __

12. $\begin{array}{r} 7 \\ -3 \\ \hline \end{array}$

13. $\begin{array}{r} 3 \\ +5 \\ \hline \end{array}$

14. $\begin{array}{r} 5 \\ +2 \\ \hline \end{array}$

15. $\begin{array}{r} 10 \\ -7 \\ \hline \end{array}$

16. $\begin{array}{r} 1 \\ +6 \\ \hline \end{array}$

17. $\begin{array}{r} 7 \\ -5 \\ \hline \end{array}$

18. $7 - 3 =$ __

19. $\begin{array}{r} 6 \\ +2 \\ \hline \end{array}$

20. $\begin{array}{r} 7 \\ -4 \\ \hline \end{array}$

THE LOWER ADDITION AND SUBTRACTION FACTS
FACILITY EXERCISES #41

Answer the problems below. There must be **no use of fingers or any counting.**

1. $8 - 5 =$ __

2. $\begin{array}{r} 5 \\ +4 \\ \hline \end{array}$

3. $\begin{array}{r} 3 \\ +6 \\ \hline \end{array}$

4. $7 - 2 =$ __

5. $3 + 4 =$ __

6. $\begin{array}{r} 8 \\ -6 \\ \hline \end{array}$

7. $7 - 3 =$ __

8. $\begin{array}{r} 9 \\ -2 \\ \hline \end{array}$

9. $\begin{array}{r} 7 \\ -2 \\ \hline \end{array}$

10. $7 - 6 =$ __

11. $\begin{array}{r} 4 \\ +4 \\ \hline \end{array}$

12. $2 + 5 =$ __

13. $\begin{array}{r} 7 \\ -4 \\ \hline \end{array}$

14. $\begin{array}{r} 9 \\ -7 \\ \hline \end{array}$

15. $\begin{array}{r} 7 \\ -6 \\ \hline \end{array}$

16. $\begin{array}{r} 3 \\ +5 \\ \hline \end{array}$

17. $\begin{array}{r} 7 \\ -1 \\ \hline \end{array}$

18. $7 - 5 =$ __

19. $4 + 3 =$ __

20. $\begin{array}{r} 6 \\ +1 \\ \hline \end{array}$

THE LOWER ADDITION AND SUBTRACTION FACTS
FACILITY EXERCISES #42

Answer the problems below. There must be **no use of fingers or any counting.**

1. $\begin{array}{r} 4 \\ +2 \\ \hline \end{array}$

2. $\begin{array}{r} 7 \\ -3 \\ \hline \end{array}$

3. $3+3=$__

4. $4+2=$__

5. $5+1=$__

6. $\begin{array}{r} 8 \\ -4 \\ \hline \end{array}$

7. $\begin{array}{r} 3 \\ +5 \\ \hline \end{array}$

8. $6-3=$__

9. $\begin{array}{r} 7 \\ -2 \\ \hline \end{array}$

10. $6-5=$__

11. $6-4=$__

12. $\begin{array}{r} 10 \\ -6 \\ \hline \end{array}$

13. $\begin{array}{r} 6 \\ -2 \\ \hline \end{array}$

14. $\begin{array}{r} 3 \\ +3 \\ \hline \end{array}$

15. $7-5=$__

16. $1+5=$__

17. $9-7=$__

18. $\begin{array}{r} 6 \\ -1 \\ \hline \end{array}$

19. $\begin{array}{r} 10 \\ -3 \\ \hline \end{array}$

20. $2+4=$__

THE LOWER ADDITION AND SUBTRACTION FACTS
FACILITY EXERCISES #43

Answer the problems below. There must be **no use of fingers or any counting.**

1. $5 + 1 =$ __

2. $2 + 4 =$ __

3. $\begin{array}{r} 7 \\ -5 \\ \hline \end{array}$

4. $\begin{array}{r} 6 \\ -3 \\ \hline \end{array}$

5. $\begin{array}{r} 6 \\ +3 \\ \hline \end{array}$

6. $6 - 2 =$ __

7. $6 + 2 =$ __

8. $6 - 5 =$ __

9. $3 + 3 =$ __

10. $\begin{array}{r} 7 \\ -2 \\ \hline \end{array}$

11. $\begin{array}{r} 7 \\ +2 \\ \hline \end{array}$

12. $\begin{array}{r} 6 \\ -4 \\ \hline \end{array}$

13. $\begin{array}{r} 6 \\ +4 \\ \hline \end{array}$

14. $\begin{array}{r} 6 \\ -1 \\ \hline \end{array}$

15. $7 - 3 =$ __

16. $7 + 3 =$ __

17. $10 - 5 =$ __

18. $\begin{array}{r} 2 \\ +4 \\ \hline \end{array}$

19. $\begin{array}{r} 6 \\ -2 \\ \hline \end{array}$

20. $5 + 3 =$ __

THE LOWER ADDITION AND SUBTRACTION FACTS
FACILITY EXERCISES #44

Answer the problems below. There must be **no use of fingers or any counting.**

1. $6-3=$__
2. $9+1=$__
3. $4+2=$__
4. $\begin{array}{r} 7 \\ \underline{-\ 2} \end{array}$
5. $7-3=$__
6. $\begin{array}{r} 7 \\ \underline{+\ 3} \end{array}$
7. $\begin{array}{r} 1 \\ \underline{+\ 5} \end{array}$
8. $6-4=$__
9. $\begin{array}{r} 6 \\ \underline{-\ 2} \end{array}$
10. $\begin{array}{r} 7 \\ \underline{+\ 2} \end{array}$
11. $3+3=$__
12. $\begin{array}{r} 8 \\ \underline{-\ 6} \end{array}$
13. $6-5=$__
14. $\begin{array}{r} 6 \\ \underline{-\ 4} \end{array}$
15. $2+8=$__
16. $7-5=$__
17. $\begin{array}{r} 6 \\ \underline{-\ 3} \end{array}$
18. $5+2=$__
19. $\begin{array}{r} 3 \\ \underline{+\ 4} \end{array}$
20. $\begin{array}{r} 6 \\ \underline{-\ 1} \end{array}$

THE LOWER ADDITION AND SUBTRACTION FACTS
FACILITY EXERCISES #45

Answer the problems below. There must be **no use of fingers or any counting.**

1. $2 + 4 =$ __
2. $\begin{array}{r} 10 \\ \underline{-\ 6} \end{array}$
3. $\begin{array}{r} 6 \\ \underline{-\ 1} \end{array}$
4. $9 - 7 =$ __
5. $1 + 5 =$ __
6. $7 - 5 =$ __
7. $\begin{array}{r} 3 \\ \underline{+\ 3} \end{array}$
8. $\begin{array}{r} 6 \\ \underline{-\ 2} \end{array}$
9. $\begin{array}{r} 10 \\ \underline{-\ 6} \end{array}$
10. $6 - 4 =$ __
11. $\begin{array}{r} 6 \\ \underline{-\ 5} \end{array}$
12. $7 - 2 =$ __
13. $4 + 2 =$ __
14. $\begin{array}{r} 7 \\ \underline{-\ 3} \end{array}$
15. $\begin{array}{r} 6 \\ \underline{+\ 3} \end{array}$
16. $\begin{array}{r} 2 \\ \underline{+\ 4} \end{array}$
17. $5 + 1 =$ __
18. $\begin{array}{r} 8 \\ \underline{-\ 4} \end{array}$
19. $3 + 5 =$ __
20. $6 - 3 =$ __

THE LOWER ADDITION AND SUBTRACTION FACTS
FACILITY EXERCISES #46

Answer the problems below. There must be **no use of fingers or any counting.**

1. $\begin{array}{r} 5 \\ +3 \\ \hline \end{array}$

2. $6 - 2 =$ ___

3. $2 + 4 =$ ___

4. $\begin{array}{r} 10 \\ -5 \\ \hline \end{array}$

5. $6 - 4 =$ ___

6. $\begin{array}{r} 7 \\ +2 \\ \hline \end{array}$

7. $7 - 2 =$ ___

8. $\begin{array}{r} 3 \\ +3 \\ \hline \end{array}$

9. $6 + 4 =$ ___

10. $6 - 1 =$ ___

11. $\begin{array}{r} 7 \\ -3 \\ \hline \end{array}$

12. $\begin{array}{r} 7 \\ +3 \\ \hline \end{array}$

13. $\begin{array}{r} 5 \\ +1 \\ \hline \end{array}$

14. $\begin{array}{r} 6 \\ -2 \\ \hline \end{array}$

15. $8 + 2 =$ ___

16. $\begin{array}{r} 6 \\ -5 \\ \hline \end{array}$

17. $4 + 2 =$ ___

18. $\begin{array}{r} 7 \\ -5 \\ \hline \end{array}$

19. $\begin{array}{r} 6 \\ -3 \\ \hline \end{array}$

20. $6 + 3 =$ ___

THE LOWER ADDITION AND SUBTRACTION FACTS
FACILITY EXERCISES #47

Answer the problems below. There must be **no use of fingers or any counting**.

1. $8 - 6 = \_\_$
2. $\begin{array}{r} 6 \\ \underline{-2} \end{array}$
3. $7 - 3 = \_\_$
4. $9 - 6 = \_\_$
5. $\begin{array}{r} 9 \\ \underline{-4} \end{array}$
6. $7 - 6 = \_\_$
7. $\begin{array}{r} 8 \\ \underline{-5} \end{array}$
8. $2 + 4 = \_\_$
9. $2 + 5 = \_\_$
10. $\begin{array}{r} 6 \\ \underline{+1} \end{array}$
11. $3 + 4 = \_\_$
12. $\begin{array}{r} 6 \\ \underline{-5} \end{array}$
13. $\begin{array}{r} 6 \\ \underline{-4} \end{array}$
14. $\begin{array}{r} 9 \\ \underline{-2} \end{array}$
15. $\begin{array}{r} 8 \\ \underline{-2} \end{array}$
16. $\begin{array}{r} 3 \\ \underline{+6} \end{array}$
17. $\begin{array}{r} 6 \\ \underline{-3} \end{array}$
18. $\begin{array}{r} 4 \\ \underline{+3} \end{array}$
19. $6 - 5 = \_\_$
20. $\begin{array}{r} 6 \\ \underline{-1} \end{array}$

THE LOWER ADDITION AND SUBTRACTION FACTS
FACILITY EXERCISES #48

Answer the problems below. There must be **no use of fingers or any counting.**

1. $\begin{array}{r} 5 \\ +3 \\ \hline \end{array}$

2. $6 - 2 =$ __

3. $2 + 4 =$ __

4. $\begin{array}{r} 10 \\ -5 \\ \hline \end{array}$

5. $6 - 4 =$ __

6. $7 + 2 =$ __

7. $7 - 2 =$ __

8. $\begin{array}{r} 3 \\ +3 \\ \hline \end{array}$

9. $\begin{array}{r} 7 \\ +3 \\ \hline \end{array}$

10. $\begin{array}{r} 7 \\ -3 \\ \hline \end{array}$

11. $6 - 1 =$ __

12. $6 + 4 =$ __

13. $\begin{array}{r} 6 \\ -5 \\ \hline \end{array}$

14. $\begin{array}{r} 6 \\ +2 \\ \hline \end{array}$

15. $\begin{array}{r} 6 \\ -2 \\ \hline \end{array}$

16. $3 + 6 =$ __

17. $\begin{array}{r} 5 \\ +1 \\ \hline \end{array}$

18. $\begin{array}{r} 2 \\ +4 \\ \hline \end{array}$

19. $7 - 5 =$ __

20. $6 - 3 =$ __

THE LOWER ADDITION AND SUBTRACTION FACTS
FACILITY EXERCISES #49

Answer the problems below. There must be **no use of fingers or any counting.**

1. $7-3=$__

2. $\begin{array}{r} 8 \\ -5 \\ \hline \end{array}$

3. $6-4=$__

4. $\begin{array}{r} 5 \\ +4 \\ \hline \end{array}$

5. $\begin{array}{r} 6 \\ -5 \\ \hline \end{array}$

6. $\begin{array}{r} 10 \\ -4 \\ \hline \end{array}$

7. $9-3=$__

8. $\begin{array}{r} 5 \\ +1 \\ \hline \end{array}$

9. $7-2=$__

10. $4+2=$__

11. $\begin{array}{r} 8 \\ -3 \\ \hline \end{array}$

12. $6-1=$__

13. $3+3=$__

14. $7-5=$__

15. $\begin{array}{r} 6 \\ +2 \\ \hline \end{array}$

16. $\begin{array}{r} 6 \\ -2 \\ \hline \end{array}$

17. $\begin{array}{r} 6 \\ +4 \\ \hline \end{array}$

18. $\begin{array}{r} 6 \\ -4 \\ \hline \end{array}$

19. $8-2=$__

20. $6-3=$__

THE LOWER ADDITION AND SUBTRACTION FACTS
FACILITY EXERCISES #50

Answer the problems below. There must be **no use of fingers or any counting.**

1. $3 + 2 =$ __ 2. $2 + 2 =$ __ 3. $2 + 1 =$ __ 4. $1 + 1 =$ __

5. $4 + 1 =$ __ 6. $3 + 1 =$ __ 7. $\begin{array}{r} 5 \\ -\ 3 \\ \hline \end{array}$ 8. $\begin{array}{r} 5 \\ -\ 2 \\ \hline \end{array}$

9. $\begin{array}{r} 4 \\ -\ 2 \\ \hline \end{array}$ 10. $\begin{array}{r} 3 \\ -\ 1 \\ \hline \end{array}$ 11. $\begin{array}{r} 5 \\ -\ 4 \\ \hline \end{array}$ 12. $\begin{array}{r} 4 \\ -\ 1 \\ \hline \end{array}$

13. $\begin{array}{r} 2 \\ -\ 1 \\ \hline \end{array}$ 14. $4 - 3 =$ __ 15. $5 - 2 =$ __ 16. $5 - 1 =$ __

17. $2 + 3 =$ __ 18. $\begin{array}{r} 2 \\ +\ 2 \\ \hline \end{array}$ 19. $5 - 3 =$ __ 20. $\begin{array}{r} 3 \\ -\ 2 \\ \hline \end{array}$

THE LOWER ADDITION AND SUBTRACTION FACTS
FACILITY EXERCISES #51

Answer the problems below. There must be **no use of fingers or any counting.**

1. $2 + 3 =$ __
2. $\begin{array}{r} 4 \\ -\ 3 \\ \hline \end{array}$
3. $6 - 2 =$ __
4. $\begin{array}{r} 5 \\ -\ 3 \\ \hline \end{array}$
5. $\begin{array}{r} 4 \\ +\ 1 \\ \hline \end{array}$
6. $\begin{array}{r} 4 \\ +\ 3 \\ \hline \end{array}$
7. $3 + 2 =$ __
8. $\begin{array}{r} 4 \\ -\ 2 \\ \hline \end{array}$
9. $\begin{array}{r} 4 \\ +\ 2 \\ \hline \end{array}$
10. $5 - 4 =$ __
11. $\begin{array}{r} 7 \\ -\ 2 \\ \hline \end{array}$
12. $\begin{array}{r} 7 \\ +\ 2 \\ \hline \end{array}$
13. $\begin{array}{r} 9 \\ -\ 6 \\ \hline \end{array}$
14. $5 - 2 =$ __
15. $5 + 2 =$ __
16. $4 - 1 =$ __
17. $2 + 2 =$ __
18. $\begin{array}{r} 8 \\ -\ 5 \\ \hline \end{array}$
19. $9 - 4 =$ __
20. $\begin{array}{r} 5 \\ -\ 1 \\ \hline \end{array}$

THE LOWER ADDITION AND SUBTRACTION FACTS
FACILITY EXERCISES #52

Answer the problems below. There must be **no use of fingers or any counting.**

1. $4 - 2 =$ __
2. $4 + 2 =$ __
3. $\begin{array}{r} 5 \\ \underline{-\ 3} \end{array}$
4. $\begin{array}{r} 5 \\ \underline{+\ 3} \end{array}$
5. $\begin{array}{r} 4 \\ \underline{+\ 3} \end{array}$
6. $4 - 3 =$ __
7. $5 + 1 =$ __
8. $5 - 1 =$ __
9. $5 - 2 =$ __
10. $\begin{array}{r} 5 \\ \underline{+\ 2} \end{array}$
11. $8 - 4 =$ __
12. $\begin{array}{r} 9 \\ \underline{-\ 7} \end{array}$
13. $7 + 3 =$ __
14. $2 - 1 =$ __
15. $\begin{array}{r} 3 \\ \underline{-\ 2} \end{array}$
16. $\begin{array}{r} 3 \\ \underline{+\ 2} \end{array}$
17. $\begin{array}{r} 4 \\ \underline{-\ 1} \end{array}$
18. $4 + 1 =$ __
19. $10 - 8 =$ __
20. $\begin{array}{r} 5 \\ \underline{+\ 4} \end{array}$

THE LOWER ADDITION AND SUBTRACTION FACTS
FACILITY EXERCISES #53

Answer the problems below. There must be **no use of fingers or any counting**.

1. $\begin{array}{r} 5 \\ -4 \\ \hline \end{array}$

2. $5 + 4 =$ __

3. $6 - 2 =$ __

4. $\begin{array}{r} 6 \\ +2 \\ \hline \end{array}$

5. $4 - 3 =$ __

6. $\begin{array}{r} 4 \\ +3 \\ \hline \end{array}$

7. $\begin{array}{r} 4 \\ -2 \\ \hline \end{array}$

8. $4 + 2 =$ __

9. $\begin{array}{r} 5 \\ -2 \\ \hline \end{array}$

10. $5 + 2 =$ __

11. $\begin{array}{r} 3 \\ +2 \\ \hline \end{array}$

12. $\begin{array}{r} 5 \\ -3 \\ \hline \end{array}$

13. $\begin{array}{r} 7 \\ -3 \\ \hline \end{array}$

14. $10 - 9 =$ __

15. $\begin{array}{r} 3 \\ -2 \\ \hline \end{array}$

16. $\begin{array}{r} 9 \\ -6 \\ \hline \end{array}$

17. $8 - 6 =$ __

18. $\begin{array}{r} 2 \\ +1 \\ \hline \end{array}$

19. $\begin{array}{r} 10 \\ -4 \\ \hline \end{array}$

20. $6 + 1 =$ __

THE LOWER ADDITION AND SUBTRACTION FACTS
FACILITY EXERCISES #54

Answer the problems below. There must be **no use of fingers or any counting.**

1. $2+1=$___
2. $2-1=$___
3. $\begin{array}{r} 4 \\ +1 \\ \hline \end{array}$
4. $\begin{array}{r} 4 \\ -1 \\ \hline \end{array}$
5. $6-4=$___
6. $6+4=$___
7. $\begin{array}{r} 4 \\ +2 \\ \hline \end{array}$
8. $\begin{array}{r} 4 \\ -2 \\ \hline \end{array}$
9. $\begin{array}{r} 3 \\ +4 \\ \hline \end{array}$
10. $5-3=$___
11. $5+3=$___
12. $3-2=$___
13. $\begin{array}{r} 3 \\ -1 \\ \hline \end{array}$
14. $\begin{array}{r} 5 \\ +1 \\ \hline \end{array}$
15. $1+1=$___
16. $\begin{array}{r} 4 \\ +4 \\ \hline \end{array}$
17. $\begin{array}{r} 7 \\ -2 \\ \hline \end{array}$
18. $\begin{array}{r} 7 \\ +2 \\ \hline \end{array}$
19. $\begin{array}{r} 6 \\ +2 \\ \hline \end{array}$
20. $\begin{array}{r} 6 \\ -2 \\ \hline \end{array}$

THE LOWER ADDITION AND SUBTRACTION FACTS
FACILITY EXERCISES #55

Answer the problems below. There must be **no use of fingers or any counting**.

1. $5 - 3$

2. $5 + 3$

3. $4 + 2$

4. $4 - 2$

5. $7 + 1 =$ __

6. $8 - 7 =$ __

7. $6 - 2 =$ __

8. $5 - 4$

9. $5 + 4$

10. $7 - 4$

11. $4 - 1 =$ __

12. $3 + 2 =$ __

13. $7 - 1 =$ __

14. $8 - 6$

15. $5 - 2 =$ __

16. $2 + 1 =$ __

17. $10 - 8$

18. $5 - 2$

19. $9 - 7 =$ __

20. $5 + 5 =$ __

THE LOWER ADDITION AND SUBTRACTION FACTS
FACILITY EXERCISES #56

Answer the problems below. There must be **no use of fingers or any counting.**

1. $5 - 2 =$ __
2. $\begin{array}{r} 2 \\ +2 \\ \hline \end{array}$
3. $7 - 2 =$ __
4. $\begin{array}{r} 5 \\ -1 \\ \hline \end{array}$
5. $5 + 1 =$ __
6. $\begin{array}{r} 8 \\ -3 \\ \hline \end{array}$
7. $\begin{array}{r} 6 \\ -4 \\ \hline \end{array}$
8. $2 + 3 =$ __
9. $4 + 5 =$ __
10. $\begin{array}{r} 1 \\ +4 \\ \hline \end{array}$
11. $10 - 8 =$ __
12. $3 - 2 =$ __
13. $\begin{array}{r} 8 \\ -2 \\ \hline \end{array}$
14. $7 - 5 =$ __
15. $\begin{array}{r} 5 \\ -3 \\ \hline \end{array}$
16. $\begin{array}{r} 9 \\ -1 \\ \hline \end{array}$
17. $2 + 8 =$ __
18. $\begin{array}{r} 3 \\ -1 \\ \hline \end{array}$
19. $1 + 2 =$ __
20. $4 - 2 =$ __

THE LOWER ADDITION AND SUBTRACTION FACTS
FACILITY EXERCISES #57

Answer the problems below. There must be **no use of fingers or any counting**.

1. $\begin{array}{r} 1 \\ +4 \\ \hline \end{array}$

2. $\begin{array}{r} 3 \\ +1 \\ \hline \end{array}$

3. $3 + 2 =$ __

4. $\begin{array}{r} 7 \\ -6 \\ \hline \end{array}$

5. $\begin{array}{r} 8 \\ -4 \\ \hline \end{array}$

6. $10 - 7 =$ __

7. $5 - 4 =$ __

8. $4 + 6 =$ __

9. $9 - 3 =$ __

10. $\begin{array}{r} 5 \\ -2 \\ \hline \end{array}$

11. $4 - 3 =$ __

12. $\begin{array}{r} 3 \\ +4 \\ \hline \end{array}$

13. $6 - 2 =$ __

14. $1 + 6 =$ __

15. $\begin{array}{r} 9 \\ -3 \\ \hline \end{array}$

16. $5 - 1 =$ __

17. $\begin{array}{r} 5 \\ -3 \\ \hline \end{array}$

18. $\begin{array}{r} 9 \\ -7 \\ \hline \end{array}$

19. $4 + 2 =$ __

20. $\begin{array}{r} 3 \\ +7 \\ \hline \end{array}$

COUNTING TO 100 AND BEYOND
FACILITY EXERCISES #58

Do each of the exercises below without looking at any charts.

1. Start at 23 and count to 50.
2. Start at 79 and count to 95.
3. Start at 15 and count to 37.
4. Start at 30 and count to 60.
5. Start at 56 and count backward to 39.
6. Start at 45 and count to 79.
7. Start at 100 and count backward to 39.
8. Start at 63 and count to 100.
9. Start at 19 and count backward to 7.
10. Start at 33 and count to 73.
11. Start at 22 and count to 32.
12. Start at 25 and count backward to 15.
13. Start at 4 and count to 14.
14. Start at 7 and count to 17.
15. Start at 16 and count to 26.
16. Start at 37 and count to 47.
17. Start at 29 and count backward to 19.
18. Start at 24 and count backward to 14.
19. Start at 89 and count to 99.
20. Start at 52 and count backward to 11.
21. Start at 66 and count backward to 46.
22. Start at 23 and count backward to 13.
23. Start at 17 and count backward to 7.
24. Start at 2 and count to 12.
25. Start at 6 and count to 16.
26. Start at 40 and count to 70.
27. Start at 38 and count backward to 18.
28. Start at 57 and count to 67.
29. Start at 76 and count to 96.
30. Start at 31 and count to 81.

COUNTING TO 100 AND BEYOND
FACILITY EXERCISES #59

Do each of the exercises below.

1. Start at 327 and count to 358.
2. Start at 189 and count to 215.
3. Start at 403 and count to 429.
4. Start at 580 and count to 621.
5. Start at 876 and count to 891.
6. Start at 365 and count to 382.
7. Start at 979 and count to 1,000.
8. Start at 700 and count to 743.
9. Start at 611 and count to 641.
10. Start at 279 and count to 305.
11. Start at 555 and count to 575.
12. Start at 240 and count to 258.
13. Start at 863 and count to 880.
14. Start at 481 and count to 511.
15. Start at 653 and count to 673.
16. Start at 379 and count to 405.
17. Start at 213 and count to 237.
18. Start at 915 and count to 938.
19. Start at 666 and count to 689.
20. Start at 777 and count to 801.
21. Start at 111 and count to 130.
22. Start at 926 and count to 955.
23. Start at 699 and count to 715.
24. Start at 345 and count to 355.
25. Start at 888 and count to 955.

COUNTING TO 100 AND BEYOND
FACILITY EXERCISES #60

Do each of the exercises below.

1. Count from 1,001 to 1,030.
2. Count from 1,080 to 1,112.
3. Count from 1,186 to 1,208.
4. Count from 1,891 to 1,913.
5. Count from 1,987 to 2,015.
6. Count from 2,096 to 2,107.
7. Count from 2,190 to 2,210.
8. Count from 2,700 to 2,731.
9. Count from 2,981 to 3,021.
10. Count from 3,621 to 3,657.
11. Count from 3,799 to 3,817.
12. Count from 3,988 to 4,002.
13. Count from 4,003 to 4,025.
14. Count from 4,555 to 4,570.
15. Count from 4,898 to 4,909.
16. Count from 4,989 to 5,000.
17. Count from 5,010 to 5,030.
18. Count from 5,666 to 5,676.
19. Count from 5,990 to 6,019.
20. Count from 6,017 to 6,030.
21. Count from 6,777 to 6,791.
22. Count from 6,993 to 7,020.
23. Count from 7,888 to 7,894.
24. Count from 7,989 to 8,023.
25. Count from 8,222 to 8,241.
26. Count from 8,379 to 8,415.
27. Count from 8,983 to 9,001.
28. Count from 9,896 to 9,911.
29. Count from 9,950 to 9,971.
30. Count from 9,981 to 10,013.

VOCABULARY AND SYMBOLISM OF NUMERATION
FACILITY EXERCISES #61

Read each of the numerals below.

(1)	(2)	(3)	(4)	(5)
23	**35**	**60**	**78**	**99**
32	**37**	**22**	**87**	**41**
20	**25**	**33**	**65**	**31**
61	**52**	**44**	**70**	**50**

(6)	(7)	(8)	(9)	(10)
46	**40**	**10**	**30**	**95**
90	**72**	**55**	**38**	**80**
69	**66**	**36**	**83**	**84**
96	**81**	**63**	**59**	**74**

VOCABULARY AND SYMBOLISM OF NUMERATION
FACILITY EXERCISES #62

Read each of the numerals below.

(1)	(2)	(3)	(4)	(5)
382	**540**	**220**	**920**	**430**
841	**172**	**325**	**555**	**660**
170	**180**	**422**	**999**	**746**
597	**110**	**877**	**666**	**977**

(6)	(7)	(8)	(9)	(10)
699	**123**	**825**	**599**	**777**
859	**321**	**621**	**261**	**693**
770	**327**	**583**	**444**	**668**
333	**723**	**488**	**222**	**553**

(11)	(12)	(13)	(14)	(15)
995	**113**	**517**	**313**	**413**
842	**115**	**919**	**212**	**617**
936	**214**	**411**	**612**	**518**
118	**816**	**111**	**715**	**912**

VOCABULARY AND SYMBOLISM OF NUMERATION
FACILITY EXERCISES #63

Read each of the numerals below.

(1)	(2)	(3)
743,429,571,169,836	**645,123,942,873,427**	**777,777,777,777,777**
429,743,836,571,169	**427,645,123,942,873**	**123,123,123,123,123**
836,571,743,429,169	**873,427,645,123,942**	**123,456,789,123,456**

(4)	(5)	(6)
987,654,321,987,654	**385,641,982,137,874**	**912,711,412,611,312**
111,111,111,111,111	**293,479,627,731,478**	**111,222,333,444,555**
513,219,716,914,617	**676,828,393,787,434**	**666,777,888,999,111**

(7)	(8)	(9)
111,212,313,414,515	**116,226,336,446,556**	**981,881,781,681,581**
119,219,319,419,519	**592,492,392,292,192**	**111,112,113,114,115**
839,561,773,495,151	**428,674,181,523,999**	**848,757,666,575,484**

VOCABULARY AND SYMBOLISM OF NUMERATION
FACILITY EXERCISES #64

Read each of the numerals below.

(1)	(2)	(3)
802,040,100,048,000	**100,200,300,400,500**	**100,010,001,100,010**
500,003,020,001,100	**710,020,030,040,050**	**405,730,069,309,820**
730,703,073,730,703	**401,002,003,004,005**	**380,007,602,033,100**

(4)	(5)	(7)
400,180,071,032,093	**707,500,030,002,000**	**200,007,040,009,060**
811,012,013,014,015	**563,080,295,008,000**	**570,220,077,909,000**
900,009,090,900,000	**990,481,016,090,000**	**809,600,006,003,080**

VOCABULARY AND SYMBOLISM OF NUMERATION
FACILITY EXERCISES #65

Read each of the numerals below.

(1)	(2)	(3)
500,050,000,000,005	**800,080,008,000,260**	**659,000,000,000,001**
602,000,010,000,002	**304,000,000,060,000**	**100,100,000,000,000**
760,009,000,500,000	**270,010,000,000,000**	**100,000,001,000,000**

(4)	(5)	(6)
394,000,671,000,125	**674,674,000,000,674**	**100,000,000,000,000**
394,000,000,671,125	**674,000,000,674,674**	**101,001,000,000,000**
394,000,000,000,125	**674,000,674,000,674**	**101,001,010,000,000**

VOCABULARY AND SYMBOLISM OF NUMERATION
FACILITY EXERCISES #66

Read each of the numerals below.

(1)	(2)	(3)
6,847,231	8,473,593,684,102	20,000,000,000,000
23,984,195	51,056,630,100,321	200,000,000
16,247,695,283	70,000,200,000	20,000

(4)	(5)	(6)
1,000,000,000,000	780,000,030,000	509,001,001
7,000,000,000	10,000	41,000,050,050
7,000,000	37,005,500	400

(7)	(8)	(9)
784,000,000,000,000	100,000	10,000,000
93,000	6,000	100,000,000
4,444,444	1,000,000	1,000,000,000

(10)	(11)	(12)
740,609,500,201	328,194	11,000
590,000,000	600,006	99,000,090
8,000,000,080	57,800,007	129,000,000,900

ADDING TWO LARGE WHOLE NUMBERS
FACILITY EXERCISES #67

Do the additions below. After completing each example, read the answer.

(1)

$$\begin{array}{r} 437{,}061{,}584 \\ +\ 252{,}923{,}214 \\ \hline \end{array}$$

(2)

$$\begin{array}{r} 241{,}624 \\ +\ 351{,}252 \\ \hline \end{array}$$

(3)

$$\begin{array}{r} 6{,}452{,}384 \\ +\ 2{,}324{,}605 \\ \hline \end{array}$$

(4)

$$\begin{array}{r} 873{,}026{,}753{,}642 \\ +\ 123{,}053{,}132{,}347 \\ \hline \end{array}$$

(5)

$$\begin{array}{r} 476{,}058{,}163 \\ +\ 23{,}120{,}434 \\ \hline \end{array}$$

(6)	(7)	(8)
$\begin{array}{r} 541{,}695{,}287 \\ +\ 41{,}204{,}210 \\ \hline \end{array}$	$\begin{array}{r} 720{,}856 \\ +\ 130{,}132 \\ \hline \end{array}$	$\begin{array}{r} 4{,}253{,}092 \\ +\ 634{,}602 \\ \hline \end{array}$
(9)	(10)	(11)
$\begin{array}{r} 680{,}295{,}314 \\ +\ 117{,}404{,}252 \\ \hline \end{array}$	$\begin{array}{r} 323{,}647 \\ +\ 423{,}221 \\ \hline \end{array}$	$\begin{array}{r} 1{,}587{,}239 \\ +\ 401{,}540 \\ \hline \end{array}$
(12)	(13)	(14)
$\begin{array}{r} 425{,}634{,}963 \\ +\ 3{,}203{,}025 \\ \hline \end{array}$	$\begin{array}{r} 730{,}643 \\ +\ 240{,}216 \\ \hline \end{array}$	$\begin{array}{r} 38{,}251 \\ +\ 42{,}150{,}637 \\ \hline \end{array}$

(15)

$$\begin{array}{r} 1{,}945{,}327{,}504 \\ +\ 5{,}042{,}542{,}302 \\ \hline \end{array}$$

(16)

$$\begin{array}{r} 2{,}549{,}182 \\ +\ 6{,}320{,}016 \\ \hline \end{array}$$

(17)	(18)	(19)
$\begin{array}{r} 255{,}320{,}647 \\ +\ 424{,}370{,}232 \\ \hline \end{array}$	$\begin{array}{r} 435{,}423 \\ +\ 243{,}264 \\ \hline \end{array}$	$\begin{array}{r} 2{,}222{,}222 \\ +\ 1{,}234{,}567 \\ \hline \end{array}$
(20)	(21)	(22)
$\begin{array}{r} 654{,}321 \\ +\ 333{,}333 \\ \hline \end{array}$	$\begin{array}{r} 652{,}704 \\ +\ 324{,}194 \\ \hline \end{array}$	$\begin{array}{r} 12{,}345 \\ +\ 44{,}444 \\ \hline \end{array}$
(23)	(24)	(25)
$\begin{array}{r} 43{,}210 \\ +\ \ 55{,}555 \\ \hline \end{array}$	$\begin{array}{r} 120 \\ +\ 777 \\ \hline \end{array}$	$\begin{array}{r} 6{,}666 \\ +\ \ 1{,}230 \\ \hline \end{array}$
(26)	(27)	(28)
$\begin{array}{r} 88 \\ +\ 11 \\ \hline \end{array}$	$\begin{array}{r} 462{,}857 \\ +\ 434{,}132 \\ \hline \end{array}$	$\begin{array}{r} 1{,}234{,}567 \\ +\ 7{,}654{,}321 \\ \hline \end{array}$

(29)

```
  654,321
+ 123,456
```

(30)

```
  825,617,431
+ 153,322,040
```

(31) 423,014 + 423,014	(32) 3,547 + 2,352	(33) 12,345,678 + 87,654,321
(34) 526,405,625 + 132,402,324	(35) 32,524 + 32,431	(36) 7,245,637,894 + 24,252,104
(37) 1,458,067,348 + 420,622,541	(38) 250,485 + 430,502	(39) 61,358,247 + 26,431,401
(40) 1,473,643 + 774,422,136	(41) 632,614 + 45,235	(42) 12,345 + 54,321

PLACE VALUE
FACILITY EXERCISES #68

What number does each digit in each number represent?

(1) 152,427,592,583,478 875 333,333,333 12	(2) 471,563,356 1,103,391,516 66,555 2,711,481,357,171
(3) 726,000 13,013,013,013 50,040,610 83	(4) 222,000,000,000 444,000,444,000 6,000,000 123,000,000,123
(5) 619,619,000,000 90 5 374,743,437	(6) 1 66 11,111 45,537,295,382,000
(7) 357,482,591,762,924 590,008,460,214 400 26,564	(8) 596,287 84,848 3,476 7,543
(9) 764,465 248 34,573 485,128	(10) 63,574,195 14,523,412 16,411,231 76,076,334,209
(11) 12,667,267,113 801 637,145,173,155,537 6,666	(12) 93,399 777,777,777,777 406 13

PLACE VALUE
FACILITY EXERCISES #69

Point to every digit in each numeral below and name its place value.

(1)	(2)	(3)
6,847,231 23,984,195 16,247,695,283	8,473,593,684,102 51,056,630,100,321 70,000,200,000	20,000,000,000,000 200,000,000 20,000
(4)	**(5)**	**(6)**
1,000,000,000,000 7,000,000,000 7,000,000	780,000,030,000 10,000 37,005,500	509,001,001 41,000,050,050 400
(7)	**(8)**	**(9)**
784,000,000,000,000 93,000 4,444,444	100,000 6,000 1,000,000	10,000,000 100,000,000 1,000,000,000
(10)	**(11)**	**(12)**
740,609,500,201 590,000,000 8,000,000,080	328,194 600,006 57,800,007	11,000 99,000,090 129,000,000,900

COMPARING WHOLE NUMBERS
FACILITY EXERCISES #70

Have learners answer the following questions:

1. The number 2,674,195 is in
- (a) which millions?
- (b) which hundred-thousands?
- (c) which ten-thousands?
- (d) which thousands?
- (e) which hundreds?
- (f) which tens?

2. The number 726,451 is in
- (a) which hundred-thousands?
- (b) which ten-thousands?
- (c) which thousands?
- (d) which hundreds?
- (e) which tens?

3. The number 407,143 is in
- (a) which hundred-thousands?
- (b) which ten-thousands?
- (c) which thousands?
- (d) which hundreds?
- (e) which tens?

4. The number 384,057 is in
- (a) which hundred-thousands?
- (b) which ten-thousands?
- (c) which thousands?
- (d) which hundreds?
- (e) which tens?

5. The number 551,403 is in
- (a) which hundred-thousands?
- (b) which ten-thousands?
- (c) which thousands?
- (d) which hundreds?
- (e) which tens?

6. The number 7,398,264 is in
- (a) which millions?
- (b) which hundred-thousands?
- (c) which ten-thousands?
- (d) which thousands?
- (e) which hundreds?

7. The number 5,000,670 is in
- (a) which millions?
- (b) which hundred-thousands?
- (c) which ten-thousands?
- (d) which thousands?
- (e) which hundreds?
- (f) which tens?

8. The number 9,002,083 is in
- (a) which millions?
- (b) which hundred-thousands?
- (c) which ten-thousands?
- (d) which thousands?
- (e) which hundreds?
- (f) which tens?

9. The number 6,050,000 is in
- (a) which millions?
- (b) which hundred-thousands?
- (c) which ten-thousands?
- (d) which thousands?
- (e) which hundreds?
- (f) which tens?

10. The number 3,600,001 is in
- (a) which millions?
- (b) which hundred-thousands?
- (c) which ten-thousands?
- (d) which thousands?
- (e) which hundreds?
- (f) which tens?

COMPARING WHOLE NUMBERS
FACILITY EXERCISES #71

For each pair of numbers below, tell which one is larger and explain why.

(1)	(2)	(3)
37 or 39 31 or 13 239 or 199	49 or 51 141 or 114 379 or 381	42 or 24 28 or 101 3,429 or 3,432
(4) 123 or 700 488 or 607 437,999 or 438,001	**(5)** 1,001 or 998 71,595 or 71,495 600,000 or 599,999	**(6)** 372 or 331 878 or 1,223 79,989 or 81,054
(7) 1,020,000 or 1,017,777 562 or 561 889,968 or 1,000,000	**(8)** 1,849 or 234 3,988 or 5,231 67,231 or 121,213	**(9)** 1,192 or 960 216 or 190 4,567 or 4,568
(10) 531 or 399 29,043 or 34,000 1,101,030 or 40,000	**(11)** 8,746 or 8,756 488 or 607 348,219 or 348,309	**(12)** 6,951 or 6,952 6,493 or 6,793 819 or 719
(13) 9,529 or 9,524 799 or 942 578,435 or 99,879	**(14)** 899 or 900 6,959 or 8,107 793,988 or 792,111	**(15)** 7,111 or 999 1,026 or 950,001 287,496 or 287,396

(16)	(17)
23,402,516 or 7,694,876 200,969 or 300,000 896,001 or 796,002	743,051,407 or 743,051,610 493,646 or 493,647 473,058 or 373,059
(18) 625,032 or 725,031 7,999,999 or 6,000,000 1,111 or 999	**(19)** 716,004 or 516,009 9,000,000 or 8,000,087 22,222 or 18,888

SUBTRACTING ONE LARGE WHOLE NUMBER FROM ANOTHER
FACILITY EXERCISES #72

Do the subtractions below. After completing each example, read your result and check the answer.

(1) **Check**

$$\begin{array}{r} 4{,}875{,}395 \\ -\ 1{,}432{,}023 \\ \hline \end{array}$$

(2) **Check**

$$\begin{array}{r} 97{,}485{,}286 \\ -\ \ 5{,}140{,}254 \\ \hline \end{array}$$

(3) **Check**

$$\begin{array}{r} 46 \\ -\ 23 \\ \hline \end{array}$$

(4) **Check**

$$\begin{array}{r} 6{,}482{,}697 \\ -\ 3{,}241{,}594 \\ \hline \end{array}$$

(5) **Check**

$$\begin{array}{r} 9 \\ -\ 2 \\ \hline \end{array}$$

(6) **Check**

$$\begin{array}{r} 87 \\ -\ \ 5 \\ \hline \end{array}$$

(7) **Check**

$$\begin{array}{r} 574{,}965{,}947 \\ -\ 204{,}423{,}620 \\ \hline \end{array}$$

(8) **Check**

$$\begin{array}{r} 864 \\ -\ 432 \\ \hline \end{array}$$

(9) **Check**

$$\begin{array}{r} 7{,}854{,}974{,}638 \\ -\ 5{,}632{,}751{,}334 \\ \hline \end{array}$$

(10) **Check**

$$\begin{array}{r} 986 \\ -\ \ \ 2 \\ \hline \end{array}$$

(11) **Check**

```
  37,946,573,057
-  4,613,240,015
```

(12) **Check**

```
  134
-  30
```

(13) **Check**

```
  583,295,871
- 240,231,631
```

(14) **Check**

```
  177
-  43
```

(15) **Check**

```
  9,763,528,492
- 2,621,425,190
```

(16) **Check**

```
  684
- 600
```

(17) **Check**

```
  529,478,352
-   9,478,352
```

(18) **Check**

```
  46,849
-    205
```

(19) **Check**

```
  2,674,593,473
- 2,243,463,151
```

(20) **Check**

```
  4,687
- 1,607
```

(21) **Check**

```
  7,586,263,980
-    44,031,260
```

(22) **Check**

```
  8,497
- 2,322
```

(23)

Check

$$\begin{array}{r} 437{,}946{,}573{,}165 \\ -\ 104{,}613{,}240{,}142 \\ \hline \end{array}$$

(24)

Check

$$\begin{array}{r} 78 \\ -\ \ 8 \\ \hline \end{array}$$

(25)

Check

$$\begin{array}{r} 869{,}753 \\ -\ 542{,}531 \\ \hline \end{array}$$

(26)

Check

$$\begin{array}{r} 297 \\ -\ \ 60 \\ \hline \end{array}$$

(27)

Check

$$\begin{array}{r} 842{,}719 \\ -\ 842{,}719 \\ \hline \end{array}$$

(28)

Check

$$\begin{array}{r} 684 \\ -\ \ \ 3 \\ \hline \end{array}$$

(29)

Check

$$\begin{array}{r} 2{,}804{,}628 \\ -\ 1{,}402{,}314 \\ \hline \end{array}$$

(30)

Check

$$\begin{array}{r} 67 \\ -\ 24 \\ \hline \end{array}$$

(31)

Check

$$\begin{array}{r} 9{,}284{,}763 \\ -\ 3{,}234{,}361 \\ \hline \end{array}$$

(32)

Check

$$\begin{array}{r} 684 \\ -\ \ 70 \\ \hline \end{array}$$

(33)

Check

$$\begin{array}{r} 874{,}597{,}648 \\ -\ 263{,}372{,}324 \\ \hline \end{array}$$

(34)

Check

$$\begin{array}{r} 7{,}598 \\ -\ \ \ \ \ 2 \\ \hline \end{array}$$

(35)

Check

```
  487,530,647
- 262,130,647
```

(36)

Check

```
  986
-  23
```

(37)

Check

```
  294,685,786
- 144,541,454
```

(38)

Check

```
  317
- 105
```

(39)

Check

```
  529,478,695
-  16,232,531
```

(40)

Check

```
  7,598
-    53
```

(41)

Check

```
  87,465
- 81,155
```

(42)

Check

```
  297
-  60
```

(43)

Check

```
  75,896
- 42,481
```

(44)

Check

```
  7,592
-    90
```

(45)

Check

```
  62,846
- 31,423
```

(46)

Check

```
  7,592
-   500
```

(47)

Check

$$\begin{array}{r} 54{,}387 \\ -\ 50{,}387 \\ \hline \end{array}$$

(48)

Check

$$\begin{array}{r} 7{,}592 \\ -\ 2{,}090 \\ \hline \end{array}$$

(49)

Check

$$\begin{array}{r} 23{,}047 \\ -\ 23{,}047 \\ \hline \end{array}$$

(50)

Check

$$\begin{array}{r} 7{,}592 \\ -\ \ \ 590 \\ \hline \end{array}$$

(51)

Check

$$\begin{array}{r} 7{,}592 \\ -\ 7{,}500 \\ \hline \end{array}$$

(52)

Check

$$\begin{array}{r} 6{,}975 \\ -\ 6{,}970 \\ \hline \end{array}$$

(53)

Check

$$\begin{array}{r} 7{,}592 \\ -\ 7{,}000 \\ \hline \end{array}$$

(54)

Check

$$\begin{array}{r} 7{,}592 \\ -\ \ \ 502 \\ \hline \end{array}$$

FACILITY EXERCISES #73
MIXED PRACTICE

Write your answers in the spaces provided.

(1)

$7+2-3-3+1=$ __

$5-3=$ __

$9-2=$ __

(2)

$8-6=$ __

$7+3=$ __

$9-2-2-3=$ __

(3)

$6+3=$ __

$10-4+2-5=$ __

$9-2=$ __

(4)

$10-7=$ __

$1+2+1+2-4=$ __

$8-0=$ __

(5)

$6-1+5-2-3=$ __

$0+0=$ __

$3+4=$ __

(6)

$3+2+4-5=$ __

$2+6=$ __

$0-0=$ __

(7)

$$\begin{array}{r} 3{,}584{,}162 \\ +\ 2{,}314{,}527 \\ \hline \end{array}$$

(8)

$$\begin{array}{r} 84{,}795{,}143 \\ -\ 14{,}463{,}102 \\ \hline \end{array}$$

(9)

$$\begin{array}{r} 782{,}046{,}593 \\ -\ 240{,}014{,}271 \\ \hline \end{array}$$

(10)

$$\begin{array}{r} 10 \\ -\ \ 6 \\ \hline \end{array}$$

(ll)

$$\begin{array}{r} 324{,}152{,}437 \\ +\ 352{,}406{,}431 \\ \hline \end{array}$$

(12)

$$\begin{array}{r} 389 \\ -\ \ \ 6 \\ \hline \end{array}$$

(13)

$$\begin{array}{r} 0 \\ +\ 5 \\ \hline \end{array}$$

(14)

$$\begin{array}{r} 10 \\ -\ 8 \\ \hline \end{array}$$

(15)

$$\begin{array}{r} 35 \\ +\ 23 \\ \hline \end{array}$$

(16)

$$\begin{array}{r} 20{,}000 \\ +\ 60{,}739 \\ \hline \end{array}$$

(17)

$$\begin{array}{r} 60{,}739 \\ -\ 20{,}000 \\ \hline \end{array}$$

(18)

$$\begin{array}{r} 10 \\ +\ 9 \\ \hline \end{array}$$

(19)

$$\begin{array}{r} 10 \\ +\ 6 \\ \hline \end{array}$$

(20)

$$\begin{array}{r} 13 \\ -\ 3 \\ \hline \end{array}$$

(21)

$$\begin{array}{r} 15 \\ -\ 5 \\ \hline \end{array}$$

22. Start at one and count quickly by threes to ten.

23. Start at nine and count backward by twos to one.

24. Start at 483 and count to 514.

25. Read each numeral below:

(a) 47
(b) 683
(c) 305
(d) 843,297,138,462,381
(e) 803,200,420,002,020
(f) 400,000,001,000,001
(g) 2,050,000
(h) 1,000,000
(i) 15,000

26. Name the number which each digit in 257,647,524,675,426 represents.

27. Name the value of each place in the number 257,647,524,675,426.

28. The number 4,872,195 is in

(a) which millions?

(b) which hundred-thousands?

(c) which ten-thousands?

(d) which thousands?

(e) which hundreds?

(f) which tens?

(g) which ones?

29. Which is larger:

(a) 9,999 or 10,001? Why?

(b) 1,111,111 or 9,000,000? Why?

FACILITY EXERCISES #74
MIXED PRACTICE

Write your answers in the spaces provided.

(1)

$8-5+4+2-3=$ ___

$6+2=$ ___

$8+0=$ ___

(2)

$9-5=$ ___

$10-4=$ ___

$4+5-1-2-1=$ ___

(3)

$7+2=$ ___

$9-2-3-1+4=$ ___

$8-6=$ ___

(4)

$3+3=$ ___

$5+1+2-3=$ ___

$10-0=$ ___

(5)

$1+2+3+4-7=$ ___

$10-8=$ ___

$10-0=$ ___

(6)

$4+4-6-2=$ ___

$3+2=$ ___

$10-0=$ ___

$$\begin{array}{r} 4{,}958{,}724 \\ -\ 1{,}021{,}203 \\ \hline \end{array}$$

(8)

$$\begin{array}{r} 37{,}493{,}132 \\ +\ 32{,}205{,}345 \\ \hline \end{array}$$

(9)

$$\begin{array}{r} 593{,}841{,}476 \\ -\ 241{,}430{,}273 \\ \hline \end{array}$$

(10)

$$\begin{array}{r} 10 \\ -\ \ 8 \\ \hline \end{array}$$

(11)

$$\begin{array}{r} 435{,}214{,}654 \\ +\ 423{,}281{,}305 \\ \hline \end{array}$$

(12)

$$\begin{array}{r} 467 \\ -\ \ 52 \\ \hline \end{array}$$

(13)

$$\begin{array}{r} 9 \\ -\ 2 \\ \hline \end{array}$$

(14)

$$\begin{array}{r} 0 \\ +\ 0 \\ \hline \end{array}$$

(15)

$$\begin{array}{r} 426 \\ +\ 253 \\ \hline \end{array}$$

(16)

$$\begin{array}{r} 45{,}000 \\ +\ 49{,}571 \\ \hline \end{array}$$

(17)

$$\begin{array}{r} 49{,}571 \\ -\ 45{,}000 \\ \hline \end{array}$$

(18)

$$\begin{array}{r} 10 \\ +\ 2 \\ \hline \end{array}$$

(19)

$$\begin{array}{r} 4{,}958{,}723 \\ +\ 1{,}021{,}204 \\ \hline \end{array}$$

(20)

$$\begin{array}{r} 19 \\ -\ \ 9 \\ \hline \end{array}$$

(21)

$$\begin{array}{r} 14 \\ -\ \ 4 \\ \hline \end{array}$$

22. Start at two and count by fours to ten.

23. Start at ten and count backwards by twos to zero.

24. Start at 89 and count to 123.

25. Read each numeral below:

(a) 60

(b) 582

(c) 401

(d) 673,673,673,376,376

(e) 500,030,002,605,001

(f) 500,003,000,000,000

(g) 207,001

(h) 1,005,000

(i) 10,000

26. Name the number which each digit in 643,364,436,182,218 represents.

27. Name the value of each place in the number 643,364,436,182,218.

28. The number 693,415 is in

(a) which hundred-thousands?

(b) which ten-thousands?

(c) which thousands?

(d) which hundreds?

(e) which tens?

(f) which ones?

29. Which is smaller:

(a) 6,000,000 or 5,999,999? Why?

(b) 200,000 or 1,999,999? Why?

FACILITY EXERCISES #75
MIXED PRACTICE

Write your answers in the spaces provided.

(1)

$6 - 2 - 1 - 3 + 4 =$ ___

$5 + 5 - 1 =$ ___

$7 - 7 =$ ___

(2)

$7 + 2 =$ ___

$7 - 2 =$ ___

$8 - 3 - 3 + 2 - 1 =$ ___

(3)

$3 + 3 + 3 - 4 =$ ___

$10 - 5 =$ ___

$5 + 4 =$ ___

(4)

$$\begin{array}{r} 8{,}154{,}247 \\ +\quad 623{,}450 \\ \hline \end{array}$$

(5)

$$\begin{array}{r} 647{,}398{,}431 \\ -\ 635{,}148{,}201 \\ \hline \end{array}$$

(6)

$$\begin{array}{r} 546{,}345{,}827 \\ +\ 321{,}352{,}062 \\ \hline \end{array}$$

(7)

$$\begin{array}{r} 7{,}967 \\ -\quad 243 \\ \hline \end{array}$$

(8)

$$\begin{array}{r} 89 \\ -\ 26 \\ \hline \end{array}$$

(9)

$$\begin{array}{r} 16{,}425{,}625 \\ +\ \ 2{,}354{,}232 \\ \hline \end{array}$$

(10)

$$\begin{array}{r} 10 \\ +\ \ 1 \\ \hline \end{array}$$

(11)

$$\begin{array}{r} 18 \\ -\ \ 8 \\ \hline \end{array}$$

(12)

$$\begin{array}{r} 3 \\ +\ 10 \\ \hline \end{array}$$

13. Start at one and count by twos to nine.

14. Start at ten and count backwards by twos to zero.

15. Start at 981 and count to 1,017.

16. Read each numeral below:

(a) 71

(b) 917

(c) 666,666,666,666,666

(d) 7,050,100,002,001

(e) 70,000,000,000,010

(f) 1,000,000

(g) 101

(h) 10,001,100

17. Name the number which each digit in 904,270,685,171 represents.

18. Name the value of each place in the number 904,270,685,171.

19. The number 61,504 is in

(a) which ten-thousands?

(b) which thousands?

(c) which hundreds?

(d) which tens?

(e) which ones?

20. Which is larger:

(a) 79 or 102? Why?

(b) 721 or 689? Why?

FACILITY EXERCISES #76
MIXED PRACTICE

Write your answers in the spaces provided.

(1)

$10 - 2 - 2 - 4 + 3 =$ ___

$4 + 4 - 1 =$ ___

$8 - 2 =$ ___

(2)

$5 - 3 =$ ___

$7 + 2 =$ ___

$6 - 3 - 3 + 2 + 2 =$ ___

(3)

$2 + 3 + 4 - 1 - 3 =$ ___

$10 - 7 - 1 =$ ___

$3 + 6 =$ ___

(4)

$$\begin{array}{r} 538{,}147{,}043 \\ -\ 430{,}126{,}032 \\ \hline \end{array}$$

(5)

$$\begin{array}{r} 438{,}203{,}164 \\ +\ 451{,}203{,}632 \\ \hline \end{array}$$

(6)

$$\begin{array}{r} 56{,}029{,}043 \\ +\ 31{,}040{,}024 \\ \hline \end{array}$$

(7)

$$\begin{array}{r} 78{,}405{,}908 \\ -\ 78{,}302{,}405 \\ \hline \end{array}$$

(8)

$$\begin{array}{r} 48 \\ -\ 28 \\ \hline \end{array}$$

(9)

$$\begin{array}{r} 18{,}407 \\ -\ 8{,}003 \\ \hline \end{array}$$

(10)

$$\begin{array}{r} 5 \\ +\ 10 \\ \hline \end{array}$$

(11)

$$\begin{array}{r} 13 \\ -\ 3 \\ \hline \end{array}$$

(12)

$$\begin{array}{r} 10 \\ +\ 4 \\ \hline \end{array}$$

13. Start at zero and count by threes to nine.

14. Start at eight and count backwards by threes to two.

15. Start at 979 and count to 1,020.

16. Read each numeral below:

(a) 6,010,101

(b) 70,101,010

(c) 1,000,000,000

(d) 20,002

17. Name the number which each digit in 843,105,690 represents.

18. Name the value of each place in the number 843,105,690.

19. The number 6,040,530 is in

(a) which millions?

(b) which hundred-thousands?

(c) which ten-thousands?

(d) which thousands?

(e) which hundreds?

(f) which tens?

(g) which ones?

20. Which is smaller:

(a) 3,000,000,000 or 29,999,999,999? Why?

(b) 9–8 or 1+1? Why?

(c) 1,000,000,000,000 or 999,999,999,999? Why?

EQUIVALENT NUMERALS
FACILITY EXERCISES #77

In the nine spaces to the right of each numeral, write nine different names for that number.

(1)

5 = __________ = __________ = __________
= __________ = __________ = __________
= __________ = __________ = __________

(2)

3 = __________ = __________ = __________
= __________ = __________ = __________
= __________ = __________ = __________

(3)

9 = __________ = __________ = __________
= __________ = __________ = __________
= __________ = __________ = __________

(4)

1 = __________ = __________ = __________
= __________ = __________ = __________
= __________ = __________ = __________

(5)

7 = __________ = __________ = __________
= __________ = __________ = __________
= __________ = __________ = __________

EQUIVALENT NUMERALS
FACILITY EXERCISES #78

Explain how you would make each transformation.

1. How can we transform 3+2 to 1+4?
2. How can we transform 6+1 to 5+2?
3. How can we transform 2+5 to 4+3?
4. How can we transform 5+4 to 2+7?
5. How can we transform 2+1 to 1+2?
6. How can we transform 8+6 to 5+9?
7. How can we transform 1+9 to 5+5?
8. How can we transform 9+7 to 10+6?
9. How can we transform 3+3 to 4+2?
10. How can we transform 2+6 to 6+2?
11. How can we transform 5+3 to 8+0?
12. How can we transform 8+6 to 10+4?
13. How can we transform 6+0 to 3+3?
14. How can we transform 9+8 to 10+7?
15. How can we transform 4+5 to 9+0?
16. How can we transform 10+0 to 4+6?
17. How can we transform 9+3 to 10+2?
18. How can we transform 8+1 to 0+9?
19. How can we transform 9+6 to 10+5?
20. How can we transform 6+6 to 10+2?
21. How can we transform 4+8 to 6+6?
22. How can we transform 9+3 to 4+8?
23. How can we transform 10+2 to 4+8?
24. How can we transform 10+2 to 9+3?
25. How can we transform 5+7 to 10+2?
26. How can we transform 6+5 to 3+8?
27. How can we transform 2+9 to 6+5?
28. How can we transform 9+2 to 7+4?
29. How can we transform 6+5 to 10+1?
30. How can we transform 10+1 to 7+4?
31. How can we transform 8+3 to 10+1?
32. How can we transform 10+1 to 2+9?
33. How can we transform 6+7 to 7+6?
34. How can we transform 4+9 to 6+7?
35. How can we transform 6+7 to 10+3?
36. How can we transform 8+5 to 9+4?
37. How can we transform 9+4 to 10+3?
38. How can we transform 8+5 to 10+3?
39. How can we transform 10+3 to 8+5?
40. How can we transform 7+7 to 10+4?

THE TEN–PLUS ADDITION FACTS
FACILITY EXERCISES #79

Write your answers in the spaces provided. **There must be no counting.**

(1)

10 + 6 = ___
6 + 10 = ___
16 – 6 = ___
16 – 10 = ___

(2)

10 + 1 = ___
1 + 10 = ___
11 – 10 = ___
11 – 1 = ___

(3)

10 + 9 = ___
9 + 10 = ___
19 – 9 = ___
19 – 10 = ___

(4)

10 + 3 = ___
3 + 10 = ___
13 – 10 = ___
13 – 3 = ___

(5)

10 + 7 = ___
17 – 10 = ___
7 + 10 = ___
17 – 7 = ___

(6)

10 + 4 = ___
14 – 4 = ___
14 – 10 = ___
4 + 10 = ___

(7)

2 + 10 = ___
10 + 2 = ___
12 – 10 = ___
12 – 2 = ___

(8)

10 + 8 = ___
18 – 10 = ___
8 + 10 = ___
18 – 8 = ___

(9)

5 + 10 = ___
10 + 5 = ___
15 – 5 = ___
15 – 10 = ___

KNOWING "NINE–PLUS," "EIGHT-PLUS," "SEVEN–PLUS," "SIX–PLUS," ADDITION FACTS WITHOUT MEMORIZATION FACILITY EXERCISES #80

Write your answers in the spaces provided. **There must be no counting.**

(1)	(2)	(3)
9 + 5 = ___	8 + 7 = ___	7 + 6 = ___
5 + 9 = ___	7 + 8 = ___	13 – 6 = ___
14 – 9 = ___	15 – 7 = ___	6 + 7 = ___
14 – 5 = ___	15 – 8 = ___	13 – 7 = ___

(4)	(5)	(6)
9 + 3 = ___	8 + 6 = ___	7 + 7 = ___
12 – 9 = ___	6 + 8 = ___	14 – 7 = ___
12 – 3 = ___	14 – 6 = ___	9 + 9 = ___
3 + 9 = ___	14 – 8 = ___	18 – 9 = ___

(7)	(8)	(9)
9 + 8 = ___	7 + 5 = ___	8 + 5 = ___
8 + 9 = ___	12 – 7 = ___	13 – 5 = ___
17 – 8 = ___	5 + 7 = ___	13 – 8 = ___
17 – 9 = ___	12 – 5 = ___	5 + 8 = ___

(10)	(11)	(12)
9 + 7 = ___	7 + 4 = ___	9 + 6 = ___
16 – 9 = ___	4 + 7 = ___	15 – 9 = ___
16 – 7 = ___	11 – 7 = ___	15 – 6 = ___
7 + 9 = ___	11 – 4 = ___	6 + 9 = ___

(13)	(14)	(15)
8 + 4 = ___	9 + 2 = ___	6 + 6 = ___
4 + 8 = ___	11 – 2 = ___	12 – 6 = ___
12 – 8 = ___	11 – 9 = ___	8 + 8 = ___
12 – 4 = ___	2 + 9 = ___	16 – 8 = ___

(16)	(17)	(18)
8 + 3 = ___	9 + 4 = ___	6 + 5 = ___
11 – 8 = ___	13 – 4 = ___	5 + 6 = ___
11 – 3 = ___	4 + 9 = ___	11 – 5 = ___
3 + 8 = ___	13 – 9 = ___	11 – 6 = ___

KNOWING "NINE–PLUS," "EIGHT-PLUS," "SEVEN–PLUS," "SIX–PLUS," ADDITION FACTS WITHOUT MEMORIZATION
FACILITY EXERCISES #81

Write your answers in the spaces provided. **There must be no counting.**

1. $9 + 8 = \_\_$
2. $\begin{array}{r} 8 \\ +3 \\ \hline \end{array}$
3. $7 + 5 = \_\_$
4. $\begin{array}{r} 6 \\ +6 \\ \hline \end{array}$
5. $\begin{array}{r} 7 \\ +7 \\ \hline \end{array}$
6. $\begin{array}{r} 9 \\ +6 \\ \hline \end{array}$
7. $8 + 7 = \_\_$
8. $9 + 4 = \_\_$
9. $\begin{array}{r} 8 \\ +6 \\ \hline \end{array}$
10. $9 + 2 = \_\_$
11. $9 + 9 = \_\_$
12. $\begin{array}{r} 8 \\ +5 \\ \hline \end{array}$
13. $\begin{array}{r} 7 \\ +6 \\ \hline \end{array}$
14. $9 + 5 = \_\_$
15. $\begin{array}{r} 8 \\ +8 \\ \hline \end{array}$
16. $9 + 7 = \_\_$
17. $8 + 4 = \_\_$
18. $6 + 5 = \_\_$
19. $\begin{array}{r} 9 \\ +3 \\ \hline \end{array}$
20. $\begin{array}{r} 7 \\ +4 \\ \hline \end{array}$

ADDING ANY TWO WHOLE NUMBERS
FACILITY EXERCISES #82

Your teacher will lead you through the examples below.

(1)	(2)	(3)
156,279	4,754,659	27,581,682
+ 397,483	+ 1,068,627	+ 35,297,483

(4)	(5)	(6)
7,294,726	5,948,602	8,426,573
+ 4,290,691	+ 1,653,298	+ 6,248,520

(7)	(8)	(9)
49,584,625	43,815,496	23,849,763
+ 73,685,395	+ 43,012,893	+ 19,984,674

(10)	(11)	(12)
123,456,789	987,654,321	54,387,147
+ 123,456,789	+ 123,456,789	+ 38,721,985

(13)	(14)	(15)
38,721,985	2,487,565	3,529,138
+ 54,378,147	+ 3,529,138	+ 2,487,565

(16)

$$\begin{array}{r} 4,638,357 \\ +\ 1,378,346 \\ \hline \end{array}$$

(17)

$$\begin{array}{r} 5,462,784 \\ +\ 2,469,742 \\ \hline \end{array}$$

(18)

$$\begin{array}{r} 2,469,742 \\ +\ 5,462,784 \\ \hline \end{array}$$

(19)

$$\begin{array}{r} 3,355,865 \\ +\ 4,576,661 \\ \hline \end{array}$$

(20)

$$\begin{array}{r} 8,634,951 \\ +\ 6,634,656 \\ \hline \end{array}$$

(21)

$$\begin{array}{r} 5,842,362 \\ +\ 9,422,945 \\ \hline \end{array}$$

(22)

$$\begin{array}{r} 7,751,694 \\ +\ 7,537,613 \\ \hline \end{array}$$

ADDING ANY TWO WHOLE NUMBERS
FACILITY EXERCISES #83

Do the following addition examples.

(1)
$$\begin{array}{r} 384 \\ +\ \ 78 \\ \hline \end{array}$$

(2)
$$\begin{array}{r} 65 \\ +\ \ 7 \\ \hline \end{array}$$

(3)
$$\begin{array}{r} 4{,}096 \\ +\ \ 258 \\ \hline \end{array}$$

(4)
$$\begin{array}{r} 83{,}024 \\ +\ \ 5{,}981 \\ \hline \end{array}$$

(5)
$$\begin{array}{r} 2{,}516 \\ +\ 3{,}980 \\ \hline \end{array}$$

(6)
$$\begin{array}{r} 42{,}009 \\ +\ \ 5{,}047 \\ \hline \end{array}$$

(7)
$$\begin{array}{r} 485 \\ +\ \ 5 \\ \hline \end{array}$$

(8)
$$\begin{array}{r} 38 \\ +\ 38 \\ \hline \end{array}$$

(9)
$$\begin{array}{r} 8{,}476 \\ +\ \ 195 \\ \hline \end{array}$$

(10)
$$\begin{array}{r} 428{,}368 \\ +\ 597{,}124 \\ \hline \end{array}$$

(11)
$$\begin{array}{r} 263 \\ +\ \ 37 \\ \hline \end{array}$$

(12)
$$\begin{array}{r} 2{,}384{,}195 \\ +\ \ 504{,}786 \\ \hline \end{array}$$

(13)
$$\begin{array}{r} 21{,}108 \\ +\ 93{,}473 \\ \hline \end{array}$$

(14)
$$\begin{array}{r} 2{,}947 \\ +\ \ 28 \\ \hline \end{array}$$

(15)
$$\begin{array}{r} 687{,}966 \\ +\ 847{,}693 \\ \hline \end{array}$$

(16)

$$\begin{array}{r} 55{,}555 \\ +\ 55{,}555 \\ \hline \end{array}$$

(17)

$$\begin{array}{r} 24{,}198 \\ +\ 28 \\ \hline \end{array}$$

(18)

$$\begin{array}{r} 608 \\ +\ 2{,}906 \\ \hline \end{array}$$

(19)

$$\begin{array}{r} 6{,}785 \\ +1{,}586{,}785 \\ \hline \end{array}$$

(20)

$$\begin{array}{r} 15 \\ +\ 8 \\ \hline \end{array}$$

(21)

$$\begin{array}{r} 37 \\ +\ 96 \\ \hline \end{array}$$

(22)

$$\begin{array}{r} 3{,}884 \\ +\ 7{,}742 \\ \hline \end{array}$$

(23)

$$\begin{array}{r} 4{,}675 \\ +\ 6{,}951 \\ \hline \end{array}$$

(24)

$$\begin{array}{r} 680{,}473 \\ +\ 875{,}294 \\ \hline \end{array}$$

(25)

$$\begin{array}{r} 962{,}381 \\ +\ 593{,}386 \\ \hline \end{array}$$

(26)

$$\begin{array}{r} 67 \\ +\ 23{,}429 \\ \hline \end{array}$$

(27)

$$\begin{array}{r} 160{,}589 \\ +\ 39{,}420 \\ \hline \end{array}$$

(28)

$$\begin{array}{r} 64{,}894 \\ +\ 935{,}106 \\ \hline \end{array}$$

(29)

$$\begin{array}{r} 8 \\ +\ 46 \\ \hline \end{array}$$

(30)

$$\begin{array}{r} 7{,}296{,}874 \\ +\ 1{,}589{,}243 \\ \hline \end{array}$$

(31)

$$\begin{array}{r} 5{,}482{,}105 \\ +\ 4{,}027{,}508 \\ \hline \end{array}$$

(32)

$$\begin{array}{r} 671 \\ +\ 283 \\ \hline \end{array}$$

(33)

$$\begin{array}{r} 64{,}000 \\ +\ 79{,}083 \\ \hline \end{array}$$

(34)	(35)	(36)
358,746 + 641,989	4,987 + 7,984	7,050 + 4,039
(37)	**(38)**	**(39)**
683,274 + 133,544	56,724 + 63,282	23,496 + 18,055
(40)	**(41)**	**(42)**
276,195 + 483,145	54,891 + 45,109	2,864 + 678,529
(43)	**(44)**	**(45)**
82 + 94	658 + 237	39 + 9
(46)	**(47)**	**(48)**
679 + 867	105 + 87	974 + 64

EQUIVALENT NUMERALS
FACILITY EXERCISES #84

Answer "yes" or "no" to each question below:

1. Can we take 8 from 3?

2. Can we take 4 from 3?

3. Can we take 5 from 2?

4. Can we take 9 from 7?

5. Can we take 5 from 8?

6. Can we take 5 from 6?

7. Can we take 7 from 4?

8. Can we take 7 from 10?

9. Can we take 7 from 7?

10. Can we take 0 from 6?

11. Can we take 9 from 1?

12. Can we take 2 from 6?

13. Can we take 10 from 0?

14. Can we take 2 from 0?

15. Can we take 5 from 4?

16. Can we take 3 from 4?

17. Can we take 1 from 4?

18. Can we take 9 from 6?

19. Can we take 4 from 8?

20. Can we take 4 from 6?

21. Can we take 3 from 2?

22. Can we take 5 from 0?

23. Can we take 10 from 9?

24. Can we take 6 from 8?

25. Can we take 6 from 6?

26. Can we take 9 from 3?

SUBTRACTION FACTS WITHOUT MEMORIZATION
FACILITY EXERCISES #85

Find the answer to each subtraction example and check it. **There must be no counting.**

(1) **Check**

$$\begin{array}{r} 15 \\ -\ 7 \\ \hline \end{array}$$

(2) **Check**

$$\begin{array}{r} 13 \\ -\ 6 \\ \hline \end{array}$$

(3) **Check**

$$\begin{array}{r} 11 \\ -\ 3 \\ \hline \end{array}$$

(4) **Check**

$$\begin{array}{r} 12 \\ -\ 8 \\ \hline \end{array}$$

(5) **Check**

$$\begin{array}{r} 17 \\ -\ 9 \\ \hline \end{array}$$

(6) **Check**

$$\begin{array}{r} 13 \\ -\ 7 \\ \hline \end{array}$$

(7) **Check**

$$\begin{array}{r} 11 \\ -\ 6 \\ \hline \end{array}$$

(8) **Check**

$$\begin{array}{r} 14 \\ -\ 9 \\ \hline \end{array}$$

(9)

Check

$$\begin{array}{r} 16 \\ -\ 9 \\ \hline \end{array}$$

(10)

Check

$$\begin{array}{r} 15 \\ -\ 8 \\ \hline \end{array}$$

(11)

Check

$$\begin{array}{r} 12 \\ -\ 9 \\ \hline \end{array}$$

(12)

Check

$$\begin{array}{r} 13 \\ -\ 5 \\ \hline \end{array}$$

(13)

Check

$$\begin{array}{r} 14 \\ -\ 6 \\ \hline \end{array}$$

(14)

Check

$$\begin{array}{r} 11 \\ -\ 7 \\ \hline \end{array}$$

(15)

Check

$$\begin{array}{r} 18 \\ -\ 9 \\ \hline \end{array}$$

(16)

Check

$$\begin{array}{r} 14 \\ -\ 8 \\ \hline \end{array}$$

(17)

Check

$$\begin{array}{r} 12 \\ -\ 4 \\ \hline \end{array}$$

(18)

Check

$$\begin{array}{r} 16 \\ -\ 8 \\ \hline \end{array}$$

(19) **Check**

$$\begin{array}{r} 14 \\ -\ 5 \\ \hline \end{array}$$

(20) **Check**

$$\begin{array}{r} 15 \\ -\ 9 \\ \hline \end{array}$$

(21) **Check**

$$\begin{array}{r} 16 \\ -\ 7 \\ \hline \end{array}$$

(22) **Check**

$$\begin{array}{r} 11 \\ -\ 5 \\ \hline \end{array}$$

(23) **Check**

$$\begin{array}{r} 12 \\ -\ 3 \\ \hline \end{array}$$

(24) **Check**

$$\begin{array}{r} 15 \\ -\ 6 \\ \hline \end{array}$$

(25) **Check**

$$\begin{array}{r} 13 \\ -\ 8 \\ \hline \end{array}$$

(26) **Check**

$$\begin{array}{r} 11 \\ -\ 9 \\ \hline \end{array}$$

(27) **Check**

$$\begin{array}{r} 12 \\ -\ 7 \\ \hline \end{array}$$

(28) **Check**

$$\begin{array}{r} 13 \\ -\ 4 \\ \hline \end{array}$$

(29) **Check**

$$\begin{array}{r} 11 \\ -\ 8 \\ \hline \end{array}$$

(30) **Check**

$$\begin{array}{r} 12 \\ -\ 6 \\ \hline \end{array}$$

(31) **Check**

$$\begin{array}{r} 13 \\ -\ 9 \\ \hline \end{array}$$

(32) **Check**

$$\begin{array}{r} 11 \\ -\ 4 \\ \hline \end{array}$$

(33) **Check**

$$\begin{array}{r} 14 \\ -\ 7 \\ \hline \end{array}$$

(34) **Check**

$$\begin{array}{r} 11 \\ -\ 2 \\ \hline \end{array}$$

(35) **Check**

$$\begin{array}{r} 12 \\ -\ 5 \\ \hline \end{array}$$

(36) **Check**

$$\begin{array}{r} 17 \\ -\ 8 \\ \hline \end{array}$$

SUBTRACTING ONE WHOLE NUMBER FROM ANOTHER
FACILITY EXERCISES #86

Your teacher will lead you through the examples below.

(1)	(2)	(3)
52,430,819	4,784,659	75,147,632
– 34,580,793	– 1,058,427	– 34,631,482

(4)	(5)	(6)
3,241,631	5,948,692	8,426,573
– 1,598,726	– 1,653,298	– 6,248,520

(7)	(8)	(9)
73,658,395	43,815,496	23,849,763
– 49,584,625	– 43,012,893	– 19,984,674

(10)	(11)	(12)
93,847,381	24,813,245	123,321
– 86,152,629	– 20,803,040	– 120,742

(13)	(14)	(15)
987,654,321	54,387,147	3,529,128
– 123,456,789	– 38,721,985	– 2,487,565

(16)

$$\begin{array}{r} 4{,}638{,}357 \\ -\ 1{,}378{,}346 \\ \hline \end{array}$$

(17)

$$\begin{array}{r} 723{,}563 \\ -\ 158{,}794 \\ \hline \end{array}$$

(18)

$$\begin{array}{r} 5{,}462{,}784 \\ -\ 2{,}469{,}742 \\ \hline \end{array}$$

(19)

$$\begin{array}{r} 4{,}576{,}661 \\ -\ 3{,}355{,}865 \\ \hline \end{array}$$

(20)

$$\begin{array}{r} 4{,}797{,}632 \\ -\ 3{,}134{,}894 \\ \hline \end{array}$$

(21)

$$\begin{array}{r} 6{,}283{,}953 \\ -\ 1{,}648{,}573 \\ \hline \end{array}$$

(22)

$$\begin{array}{r} 5{,}476{,}581 \\ -\ 2{,}385{,}945 \\ \hline \end{array}$$

SUBTRACTING ONE WHOLE NUMBER FROM ANOTHER
FACILITY EXERCISES #87

Find the answer to each example below and check it.

(1) **Check**

$$\begin{array}{r} 639{,}073 \\ -\ 247{,}291 \\ \hline \end{array}$$

(2) **Check**

$$\begin{array}{r} 438{,}759{,}384 \\ -\ 276{,}384{,}592 \\ \hline \end{array}$$

(3) **Check**

$$\begin{array}{r} 86 \\ -\ 28 \\ \hline \end{array}$$

(4) **Check**

$$\begin{array}{r} 5{,}682{,}794 \\ -\ \ \ 786{,}496 \\ \hline \end{array}$$

(5) **Check**

$$\begin{array}{r} 23 \\ -\ \ 8 \\ \hline \end{array}$$

(6) **Check**

$$\begin{array}{r} 385 \\ -\ \ \ 7 \\ \hline \end{array}$$

(7) **Check**

$$\begin{array}{r} 4{,}329{,}856 \\ -\ 1{,}485{,}278 \\ \hline \end{array}$$

(8) **Check**

$$\begin{array}{r} 428{,}140 \\ -\ 389{,}362 \\ \hline \end{array}$$

(9) **Check**

$$\begin{array}{r} 70 \\ -\ \ 6 \\ \hline \end{array}$$

(10) **Check**

$$\begin{array}{r} 2{,}384{,}073{,}374 \\ -\ \ \ \ 785{,}608{,}743 \\ \hline \end{array}$$

(11) **Check**

$$\begin{array}{r} 4{,}372{,}523 \\ -\quad 86{,}705 \\ \hline \end{array}$$

(12) **Check**

$$\begin{array}{r} 7{,}465{,}262{,}418 \\ -\quad 83{,}198{,}269 \\ \hline \end{array}$$

(13) **Check**

$$\begin{array}{r} 23 \\ -\ 7 \\ \hline \end{array}$$

(14) **Check**

$$\begin{array}{r} 526{,}076 \\ -\ 284{,}027 \\ \hline \end{array}$$

(15) **Check**

$$\begin{array}{r} 16 \\ -\ 7 \\ \hline \end{array}$$

(16) **Check**

$$\begin{array}{r} 4{,}807 \\ -\ 2{,}467 \\ \hline \end{array}$$

(17) **Check**

$$\begin{array}{r} 593 \\ -\ 479 \\ \hline \end{array}$$

(18) **Check**

$$\begin{array}{r} 7{,}705{,}522{,}879 \\ -\ 3{,}852{,}761{,}893 \\ \hline \end{array}$$

(19) **Check**

$$\begin{array}{r} 84{,}197{,}152 \\ -\ 27{,}828{,}196 \\ \hline \end{array}$$

(20) **Check**

$$\begin{array}{r} 17{,}254{,}108 \\ -\quad 8{,}627{,}504 \\ \hline \end{array}$$

(21) **Check**

$$\begin{array}{r} 43{,}816 \\ -\quad 9{,}832 \\ \hline \end{array}$$

(22) **Check**

$$\begin{array}{r} 24 \\ -\ 12 \\ \hline \end{array}$$

(23) **Check**

$$\begin{array}{r} 13{,}235 \\ -\ \ 8{,}147 \\ \hline \end{array}$$

(24) **Check**

$$\begin{array}{r} 456{,}384{,}532 \\ -\ 271{,}247{,}289 \\ \hline \end{array}$$

(25) **Check**

$$\begin{array}{r} 72 \\ -\ 36 \\ \hline \end{array}$$

(26) **Check**

$$\begin{array}{r} 9 \\ -\ 5 \\ \hline \end{array}$$

(27) **Check**

$$\begin{array}{r} 10{,}842 \\ -\ \ 10{,}459 \\ \hline \end{array}$$

(28) **Check**

$$\begin{array}{r} 94 \\ -\ 47 \\ \hline \end{array}$$

(29) **Check**

$$\begin{array}{r} 39 \\ -\ \ 5 \\ \hline \end{array}$$

(30) **Check**

$$\begin{array}{r} 456 \\ -\ \ 28 \\ \hline \end{array}$$

(31) **Check**

$$\begin{array}{r} 83{,}192 \\ -\ \ \ \ \ \ 36 \\ \hline \end{array}$$

(32) **Check**

$$\begin{array}{r} 1{,}111{,}111 \\ -\ \ \ \ 234{,}572 \\ \hline \end{array}$$

(33) **Check**

$$\begin{array}{r} 36 \\ -\ 27 \\ \hline \end{array}$$

(34) **Check**

$$\begin{array}{r} 9{,}515{,}475 \\ -\ 5{,}753{,}862 \\ \hline \end{array}$$

(35) **Check**

$$\begin{array}{r} 507 \\ -\ 253 \\ \hline \end{array}$$

(36) **Check**

$$\begin{array}{r} 9{,}422{,}945 \\ -\ 5{,}842{,}362 \\ \hline \end{array}$$

(37) **Check**

$$\begin{array}{r} 51{,}694 \\ -\ 37{,}613 \\ \hline \end{array}$$

(38) **Check**

$$\begin{array}{r} 6{,}073 \\ -\ 2{,}748 \\ \hline \end{array}$$

(39) **Check**

$$\begin{array}{r} 921 \\ -\ 583 \\ \hline \end{array}$$

(40) **Check**

$$\begin{array}{r} 9{,}706 \\ -\ 792 \\ \hline \end{array}$$

(41) **Check**

$$\begin{array}{r} 47 \\ -\ 39 \\ \hline \end{array}$$

(42) **Check**

$$\begin{array}{r} 9{,}515{,}475 \\ -\ 4{,}295{,}879 \\ \hline \end{array}$$

(43) **Check**

$$\begin{array}{r} 7{,}481 \\ -\ 6{,}854 \\ \hline \end{array}$$

(44) **Check**

$$\begin{array}{r} 9{,}275{,}123 \\ -\ 5{,}834{,}483 \\ \hline \end{array}$$

(45) **Check**

$$\begin{array}{r} 50 \\ -\ 14 \\ \hline \end{array}$$

(46) **Check**

$$\begin{array}{r} 2{,}317{,}856 \\ -\ 1{,}768{,}957 \\ \hline \end{array}$$

SUBTRACTING ONE WHOLE NUMBER FROM ANOTHER
FACILITY EXERCISES #88

Find the answer to each example below and check it.

(1)

Check

$$\begin{array}{r} 600 \\ -\quad 87 \\ \hline \end{array}$$

(2)

Check

$$\begin{array}{r} 105 \\ -\ 83 \\ \hline \end{array}$$

(3)

Check

$$\begin{array}{r} 4{,}000 \\ -\ 2{,}846 \\ \hline \end{array}$$

(4)

Check

$$\begin{array}{r} 105 \\ -\ 87 \\ \hline \end{array}$$

(5)

Check

$$\begin{array}{r} 70{,}000 \\ -\quad 6{,}589 \\ \hline \end{array}$$

(6)

Check

$$\begin{array}{r} 200{,}000 \\ -\ 153{,}476 \\ \hline \end{array}$$

(7)

Check

$$\begin{array}{r} 80 \\ -\ 36 \\ \hline \end{array}$$

(8)

Check

$$\begin{array}{r} 200{,}800 \\ -\ 153{,}976 \\ \hline \end{array}$$

(9)

Check

$$\begin{array}{r} 503 \\ -\ 276 \\ \hline \end{array}$$

(10)

Check

$$\begin{array}{r} 800{,}000 \\ -\ 360{,}425 \\ \hline \end{array}$$

(11)

Check

$$\begin{array}{r} 9{,}001 \\ -\ 6{,}753 \\ \hline \end{array}$$

(12)

Check

$$\begin{array}{r} 29{,}000 \\ -\quad 8{,}070 \\ \hline \end{array}$$

(13) **Check**

$$\begin{array}{r} 1{,}000 \\ -\ \ 678 \\ \hline \end{array}$$

(14) **Check**

$$\begin{array}{r} 111{,}000 \\ -\ \ 87{,}597 \\ \hline \end{array}$$

(15) **Check**

$$\begin{array}{r} 100 \\ -\ \ 28 \\ \hline \end{array}$$

(16) **Check**

$$\begin{array}{r} 100{,}000 \\ -\ \ 7{,}693 \\ \hline \end{array}$$

(17) **Check**

$$\begin{array}{r} 10{,}010 \\ -\ \ 9{,}456 \\ \hline \end{array}$$

(18) **Check**

$$\begin{array}{r} 101{,}010 \\ -\ \ 23{,}015 \\ \hline \end{array}$$

(19) **Check**

$$\begin{array}{r} 11{,}111 \\ -\ \ 1{,}504 \\ \hline \end{array}$$

(20) **Check**

$$\begin{array}{r} 508{,}070 \\ -\ \ 50{,}307 \\ \hline \end{array}$$

(21) **Check**

$$\begin{array}{r} 51{,}111 \\ -\ 27{,}382 \\ \hline \end{array}$$

(22) **Check**

$$\begin{array}{r} 101{,}110 \\ -\ \ 91{,}208 \\ \hline \end{array}$$

(23) **Check**

$$\begin{array}{r} 70 \\ -\ 14 \\ \hline \end{array}$$

(24) **Check**

$$\begin{array}{r} 111{,}111 \\ -\ \ 89{,}897 \\ \hline \end{array}$$

(25) **Check**

$$\begin{array}{r} 63{,}857 \\ -\ \ 3{,}000 \\ \hline \end{array}$$

(26) **Check**

$$\begin{array}{r} 101{,}010 \\ -\ \ 57{,}673 \\ \hline \end{array}$$

(27) **Check**

$$\begin{array}{r} 63{,}857 \\ -\ \ \ \ 800 \\ \hline \end{array}$$

(28) **Check**

$$\begin{array}{r} 333{,}333 \\ -\ 152{,}628 \\ \hline \end{array}$$

(29) **Check**

$$\begin{array}{r} 63{,}857 \\ -\ \ \ \ \ 50 \\ \hline \end{array}$$

(30) **Check**

$$\begin{array}{r} 111{,}111 \\ -\ \ 86{,}905 \\ \hline \end{array}$$

(31) **Check**

$$\begin{array}{r} 63{,}857 \\ -\ \ \ \ \ 50 \\ \hline \end{array}$$

(32) **Check**

$$\begin{array}{r} 61 \\ -\ \ 8 \\ \hline \end{array}$$

(33) **Check**

$$\begin{array}{r} 1{,}800 \\ -\ \ \ 321 \\ \hline \end{array}$$

(34) **Check**

$$\begin{array}{r} 5{,}119 \\ -\ 2{,}467 \\ \hline \end{array}$$

(35) **Check**

$$\begin{array}{r} 300 \\ -\ 111 \\ \hline \end{array}$$

(36) **Check**

$$\begin{array}{r} 80 \\ -\ \ 9 \\ \hline \end{array}$$

(37)

Check

$$\begin{array}{r} 570 \\ -\quad 6 \\ \hline \end{array}$$

(38)

Check

$$\begin{array}{r} 1{,}100{,}110 \\ -\quad 694{,}181 \\ \hline \end{array}$$

(39)

Check

$$\begin{array}{r} 222 \\ -\ 184 \\ \hline \end{array}$$

(40)

Check

$$\begin{array}{r} 3{,}200 \\ -\quad 84 \\ \hline \end{array}$$

(41)

Check

$$\begin{array}{r} 4{,}100 \\ -\ 3{,}000 \\ \hline \end{array}$$

(42)

Check

$$\begin{array}{r} 30{,}000 \\ -\quad 700 \\ \hline \end{array}$$

(43)

Check

$$\begin{array}{r} 7{,}100 \\ -\quad 593 \\ \hline \end{array}$$

(44)

Check

$$\begin{array}{r} 10{,}000 \\ -\quad 500 \\ \hline \end{array}$$

(45)

Check

$$\begin{array}{r} 3{,}100 \\ -\quad 517 \\ \hline \end{array}$$

(46)

Check

$$\begin{array}{r} 1{,}010 \\ -\quad 840 \\ \hline \end{array}$$

(47)

Check

$$\begin{array}{r} 1{,}010 \\ -\quad 96 \\ \hline \end{array}$$

(48)

Check

$$\begin{array}{r} 200{,}000 \\ -\quad 8{,}000 \\ \hline \end{array}$$

PREPARING TO "TELL THE TRUTH" WHEN ADDING OR SUBTRACTING WHOLE NUMBERS
FACILITY EXERCISES #89

Respond to each expression below as indicated by the example provided. The required response for #22 (below) is "Seventeen hundreds equal seven hundreds plus ten hundreds."

1. Eighteen thousands
2. Thirteen tens
3. Fourteen hundreds
4. Twelve ten-thousands
5. Nineteen tens
6. Sixteen hundreds
7. Fifteen hundred-thousands
8. Twelve millions
9. Seventeen thousands
10. Fourteen ten-thousands
11. Eighteen tens
12. Thirteen hundreds
13. Fourteen hundred-thousands
14. Twelve thousands
15. Eleven tens
16. Seventeen ten-thousands
17. Fifteen hundreds
18. Thirteen thousands
19. Sixteen tens
20. Eleven ten-thousands
21. Twelve hundred-thousands
22. Seventeen hundreds
23. Eighteen ten-thousands
24. Eleven hundred-thousands
25. Fifteen tens
26. Twelve hundreds
27. Eleven thousands
28. Nineteen ten-thousands
29. Thirteen hundred-thousands
30. Fourteen thousands

PREPARING TO "TELL THE TRUTH" WHEN ADDING OR SUBTRACTING WHOLE NUMBERS FACILITY EXERCISES #90

Answer the questions which apply to each numeral below.

"Which digit is in the ten-billions' place?"
"Which digit is in the thousands' place?"
"Which digit is in the hundred-millions' place?"
"Which digit is in the tens' place?"
"Which digit is in the hundred-thousands' place?"
"Which digit is in the ones' place?"
"Which digit is in the millions' place?"
"Which digit is in the ten-thousands' place?"
"Which digit is in the hundreds' place?"
"Which digit is in the trillions' place?"
"Which digit is in the hundred-billions' place?"
"Which digit is in the ten-millions' place?"
"Which digit is in the billions' place?"

(1)	(2)	(3)
3,704,684,592	**357,380,578**	**9,481,295,170,638**
527,681,592,134	**69,046,095**	**5,675,805**
1,525,004	**6,714,502,387,642**	**396,060,901,275**

(4)	(5)	(6)
261,234,567	**17,654,321**	**42,947,632,750**
9,315,625	**392,161,778,453**	**385,946**
75,378,423	**5,762,326,549,103**	**5,762,326,549,103**

(7)	(8)	(9)
3,653,281,594	**46,539**	**7,964,831**
4,808,494	**61,222,334**	**39,753,135**
546	**2,987,567**	**9,426**

"TELLING THE TRUTH" WHEN ADDING WHOLE NUMBERS
FACILITY EXERCISES #91

Do the following additions on the blackboard while telling the truth to your classmates with proper place value terminology.

(1)
249,846,574
+ 582,789,868

(2)
476,382,476
+ 451,965,278

(3)
83,874
+ 5,981

(4)
68,573,592
+ 65,149,283

(5)
7,843,947
+ 693,827

(6)
4,289,653
+ 1,520,277

(7)
53,872
+ 26,594

(8)
2,046,385
+ 1,053,615

(9)
4,385,716
+ 2,614,284

(10)
7,234,567
+ 1,765,433

(11)
428,368
+ 5,047

(12)
2,384,195
+ 504,786

(13)
21,108
+ 93,473

(14)
574,687,966
+ 589,847,693

(15)
680,473
+ 875,294

(16)
678,962,381
+ 238,593,386

(17)
837,296,874
+ 939,589,243

(18)
573,195
+ 427,148

SOLVING WORD PROBLEMS
FACILITY EXERCISES #92

Be sure to use only what you have learned to this point (in this book) for solving the word problems below:

1. Last year 12,987,469 people lived in the city of Oz. This year 2,576,387 more people are living there. How many people are living in Oz this year?

2. In the month of April 14,693 persons went to the college basketball game. In May 8,678 persons attended. What is the attendance over these two months?

3. Sharon has $6,307 in her savings account. How much will she have in her account if she deposits $898?

4. Two cars are on sale. One costs $16,575 and the other costs $9,463. How much would it cost to buy both cars?

5. By car it is 154 miles from Shady Grove to Pinkerton and 259 miles from Pinkerton to Humbletown. How many miles by car from Shady Grove to Humbletown?

6. At the stadium 58,463 persons attended the first playoff game and 49,987 attended the second. What is the total attendance of both games?

7. Find the cost of Mrs. Thompson's washer and dryer if the washer costs $396 and the dryer costs $289.

8. The Happytown fair lasts for two days. Attendance on the first day was 28,425 and on the second day was 19,694. What was the total attendance at the fair?

9. Mr. Allen bought a printing machine for $4,575. He had to pay a tax of $366. How much did he pay for the printing machine?

10. What is the cost of two houses at the same price if one of them costs $629,487?

11. If two calculators cost $74, what would you pay for four of them? What would you pay for six of them?

12. A number of cars cost the same amount. What is the cost of two of them if one costs $23,847? What is the cost of three of them? What would four of them cost?

13. If four machines cost $63,478, what would you pay for eight of them?

14. If 5 pens cost $185, what is the cost of 10 pens? What is the cost of 15 pens?

15. If 10 pounds of nuts cost $40, what is the cost of 15 pounds?

16. If 2 caps cost $12, what is the cost of 3 of them? What is the cost of 6 of them?

17. Six notebooks cost $18. What is the cost of 9 notebooks? What is the cost of 18 notebooks?

"TELLING THE TRUTH" WHEN SUBTRACTING WHOLE NUMBERS
FACILITY EXERCISES #93

Do the following subtractions on the blackboard while telling the truth to your classmates with proper place value terminology.

(1)
$$\begin{array}{r} 3,846,574 \\ -\ 2,789,868 \\ \hline \end{array}$$

(2)
$$\begin{array}{r} 756,382,476 \\ -\ 681,965,278 \\ \hline \end{array}$$

(3)
$$\begin{array}{r} 83,874 \\ -\ 5,981 \\ \hline \end{array}$$

(4)
$$\begin{array}{r} 4,573,582 \\ -\ 149,283 \\ \hline \end{array}$$

(5)
$$\begin{array}{r} 7,843,947 \\ -\ 693,827 \\ \hline \end{array}$$

(6)
$$\begin{array}{r} 4,289,653 \\ -\ 1,520,277 \\ \hline \end{array}$$

(7)
$$\begin{array}{r} 732,946,385 \\ -\ 451,053,615 \\ \hline \end{array}$$

(8)
$$\begin{array}{r} 4,385,716 \\ -\ 2,614,284 \\ \hline \end{array}$$

(9)
$$\begin{array}{r} 7,234,567 \\ -\ 1,765,433 \\ \hline \end{array}$$

(10)
$$\begin{array}{r} 597,124 \\ -\ 428,368 \\ \hline \end{array}$$

(11)
$$\begin{array}{r} 2,384,195 \\ -\ 504,786 \\ \hline \end{array}$$

(12)
$$\begin{array}{r} 91,108 \\ -\ 23,473 \\ \hline \end{array}$$

(13)
$$\begin{array}{r} 887,966 \\ -\ 647,693 \\ \hline \end{array}$$

(14)
$$\begin{array}{r} 875,294 \\ -\ 680,473 \\ \hline \end{array}$$

(15)
$$\begin{array}{r} 542,962,381 \\ -\ 343,593,386 \\ \hline \end{array}$$

(16)
$$\begin{array}{r} 160,589 \\ -\ 39,420 \\ \hline \end{array}$$

(17)
$$\begin{array}{r} 935,106 \\ -\ 94,894 \\ \hline \end{array}$$

(18)
$$\begin{array}{r} 7,296,874 \\ -\ 1,589,243 \\ \hline \end{array}$$

FACILITY EXERCISES #94
MIXED PRACTICE

Write your answers in the spaces provided.

1. Find answers to the examples below without use of fingers:

(a) $5 + 4 - 6 + 2 =$ ___

(b) $7 + 3 - 5 + 4 =$ ___

(c) $3 + 3 + 3 - 2 - 4 + 6 =$ ___

(d) $9 + 1 - 7 + 4 - 2 + 4 =$ ___

(e) $10 - 7 + 6 - 3 + 4 =$ ___

(f) $9 - 2 - 3 + 4 - 6 + 5 =$ ___

(g) $4 + 4 + 2 - 5 + 3 - 1 =$ ___

(h) $2 + 2 + 2 + 2 - 3 - 3 + 7 - 3 =$ ___

Be sure you check the answer to each subtraction example below.

(2)

$$\begin{array}{r} 2{,}583{,}794 \\ +\ 6{,}742{,}938 \\ \hline \end{array}$$

(3)

$$\begin{array}{r} 64{,}342{,}047 \\ -\ 37{,}819{,}564 \\ \hline \end{array}$$

(4)

$$\begin{array}{r} 267{,}453{,}623 \\ -\ \ 72{,}847{,}946 \\ \hline \end{array}$$

(5)

$$\begin{array}{r} 793{,}000{,}438 \\ -\ 298{,}462{,}736 \\ \hline \end{array}$$

(6)

$$\begin{array}{r} 59{,}407{,}086 \\ +\ 78{,}498{,}856 \\ \hline \end{array}$$

(7)

$$\begin{array}{r} 794{,}285{,}392 \\ +\ 532{,}926{,}809 \\ \hline \end{array}$$

(8)

$$\begin{array}{r} 9{,}001 \\ +\ \ \ 483 \\ \hline \end{array}$$

(9)

$$\begin{array}{r} 100{,}010 \\ -\ \ 96{,}473 \\ \hline \end{array}$$

(10)

$$\begin{array}{r} 5{,}000{,}010 \\ -\ 2{,}347{,}596 \\ \hline \end{array}$$

11. Start at two and count by fours to eighteen.

12. Start at five and count by fives to twenty.

13. Start at ten and count backwards by twos to zero.

14. Start at 887 and count to 917.

15. Read each numeral below:
- **(a)** 593,420,672,279,831
- **(b)** 70,020,250,007,300
- **(c)** 5,000,000,000,000
- **(d)** 280,056
- **(e)** 1,100,010,001

16. Name the number which each digit in 593,420,672,279,831 represents.

17. Name the value of each place in the number 593,420,672,279,831.

18. The number 7,707,070,777 is in
- **(a)** which ten-millions?
- **(b)** which thousands?
- **(c)** which ones?
- **(d)** which billions?
- **(e)** which hundred-thousands?
- **(f)** which tens?
- **(g)** which ten-thousands?
- **(h)** which hundred-millions?
- **(i)** which tens?
- **(j)** which millions?

19. Which is larger:
- **(a)** 6,482,571 or 6,482,671? Why?
- **(b)** 1,000,000,000 or 10,989,997,999? Why?

20. Tell the truth as you do the addition and subtraction examples below:

$$\begin{array}{r} 945{,}684 \\ +\ 592{,}679 \\ \hline \end{array} \qquad \begin{array}{r} 1{,}750{,}436 \\ -\ \ 964{,}587 \\ \hline \end{array}$$

FACILITY EXERCISES #95
MIXED PRACTICE

Write your answers in the spaces provided.

1. Find answers to the examples below without use of fingers:

(a) $4+3-5+3-1=$ ___

(b) $8-6+5-3+5=$ ___

(c) $10-3-4+5-2=$ ___

(d) $3+3+2+2-10=$ ___

(e) $7+2-3-2+5=$ ___

(f) $9-5-2-1+5+3=$ ___

(g) $10-2-3+5-4-2=$ ___

(h) $6-4+5-3+2=$ ___

Be sure you check the answer to each subtraction example below.

(2)
$$\begin{array}{r} 583{,}027{,}418 \\ -\ 495{,}673{,}809 \\ \hline \end{array}$$

(3)
$$\begin{array}{r} 495{,}872{,}465 \\ +\ 743{,}839{,}583 \\ \hline \end{array}$$

(4)
$$\begin{array}{r} 6{,}782{,}493 \\ +\ \quad 979{,}687 \\ \hline \end{array}$$

(5)
$$\begin{array}{r} 89{,}540{,}009 \\ -\ \quad 9{,}837{,}059 \\ \hline \end{array}$$

(6)
$$\begin{array}{r} 1{,}000{,}000 \\ -\ \qquad 50{,}627 \\ \hline \end{array}$$

(7)
$$\begin{array}{r} 497 \\ +\ 985 \\ \hline \end{array}$$

(8)
$$\begin{array}{r} 123{,}456{,}780 \\ -\ \qquad 9{,}876{,}981 \\ \hline \end{array}$$

(9)
$$\begin{array}{r} 300 \\ -\ 186 \\ \hline \end{array}$$

(10)
$$\begin{array}{r} 7{,}000{,}400{,}206 \\ -\ 2{,}305{,}642{,}184 \\ \hline \end{array}$$

11. Start at one and count by threes to ten.

12. Start at nine and count backward by threes to zero.

13. Start at 461 and count to 490.

14. Read each numeral below:
(a) 58,196,321
(b) 301,000
(c) 60,000,000,666
(d) 20,202,020

15. Name the number which each digit in 61,584,209 represents.

16. Name the value of each place in the number 61,584,209.

17. The number 20,202,020 is in
(a) which tens?
(b) which thousands?
(c) which millions?
(d) which ones?
(e) which ten-thousands?
(f) which hundreds?
(g) which ten-millions?

18. Which is smaller:
(a) 280,479 or 280,480?
(b) 10,101,010 or 10,010,010?

19. Tell the truth as you do the addition and subtraction examples below:

$$\begin{array}{r} 435{,}087 \\ -\ 198{,}598 \\ \hline \end{array} \qquad \begin{array}{r} 435{,}087 \\ +\ 198{,}598 \\ \hline \end{array}$$

FACILITY EXERCISES #96
MIXED PRACTICE

Write your answers in the spaces provided.

1. Find answers to the examples below without use of fingers:

(a)	$6 - 5 + 7 - 4 + 3 =$ ___	**(d)**	$4 + 6 - 3 + 2 - 1 - 3 =$ ___
(b)	$7 + 1 - 6 - 2 + 9 =$ ___	**(e)**	$8 - 6 + 5 - 4 + 3 =$ ___
(c)	$1 + 6 - 4 + 3 + 3 =$ ___	**(f)**	$2 + 3 + 3 - 5 + 4 =$ ___

Be sure you check the answer to each subtraction example below.

(2)	(3)	(4)
4,928,574	73,581,042	67,948
+ 3,856,275	− 6,928,576	+ 67,948

(5)	(6)	(7)
3,051	6,000,000	30,100
− 2,975	− 1,582,350	− 9,701

8. Start at two and count by threes to eight.

9. Start at ten and count backward by threes to one.

10. Start at 286 and count to 321.

11. Read each numeral below:
- **(a)** 60,000
- **(b)** 10,000,000
- **(c)** 108,100,010
- **(d)** 489,000,287,000,192

12. Name the number which each digit in 62,845 represents.

13. Name the value of each place in the number 62,845.

14. The number 3,785,493,672 is in
- **(a)** which thousands?
- **(b)** which ones?
- **(c)** which ten-millions?
- **(d)** which hundred-thousands?
- **(e)** which hundreds?
- **(f)** which millions?
- **(g)** which ten-thousands?
- **(h)** which tens?
- **(i)** which billions?
- **(j)** which hundred-millions?

15. Which is smaller:
- **(a)** 5,984,637,592 or 5,984,637,600? Why?
- **(b)** 100,000 or 99,999? Why?

16. Tell the truth as you do the addition and subtraction examples below:

$$\begin{array}{r} 1{,}748{,}396 \\ +\quad 87{,}847 \\ \hline \end{array} \qquad \begin{array}{r} 1{,}748{,}396 \\ -\quad 87{,}847 \\ \hline \end{array}$$

FACILITY EXERCISES #97
MIXED PRACTICE

Write your answers in the spaces provided.

1. Find answers to the examples below without use of fingers:

(a) $8 - 5 + 4 - 2 + 5 =$ ___ (c) $4 + 2 - 3 - 3 + 5 - 4 =$ ___

(b) $6 + 3 + 1 - 4 - 4 + 3 =$ ___ (d) $2 + 3 + 5 - 9 + 7 - 3 =$ ___

Be sure you check the answer to each subtraction example below.

(2)
$$\begin{array}{r} 49{,}073{,}621 \\ -\quad 825{,}687 \\ \hline \end{array}$$

(3)
$$\begin{array}{r} 859{,}067{,}481 \\ -\quad 79{,}436{,}795 \\ \hline \end{array}$$

(4)
$$\begin{array}{r} 859{,}067{,}481 \\ +\quad 79{,}436{,}795 \\ \hline \end{array}$$

(5)
$$\begin{array}{r} 674{,}851 \\ +\ 987{,}659 \\ \hline \end{array}$$

(6)
$$\begin{array}{r} 50{,}000 \\ -\ 27{,}003 \\ \hline \end{array}$$

(7)
$$\begin{array}{r} 80 \\ -\ 37 \\ \hline \end{array}$$

8. Start at one and count by twos to nine.

9. Start at nine and count backwards by threes to zero.

10. Start at 590 and count to 623.

11. Read each numeral below:
  (a) 313,131
  (b) 1,100
  (c) 60,060,060,060
  (d) 201,000,000,000,201

12. Name the number which each digit in 60,060,060,060 represents.

13. Name the value of each place in the number 60,060,060,060.

14. The number 5,143 is in

(a) which hundreds?

(b) which thousands?

(c) which tens?

(d) which ones?

15. Which is smaller:

(a) 54,938,000 or 54,937,999? Why?

(b) 100,000 or 99,999? Why?

16. Tell the truth as you do the addition and subtraction examples below:

$$\begin{array}{r} 4{,}582{,}734 \\ -\ 1{,}827{,}759 \\ \hline \end{array} \qquad \begin{array}{r} 4{,}582{,}734 \\ +\ 1{,}827{,}759 \\ \hline \end{array}$$

SOLVING WORD PROBLEMS
FACILITY EXERCISES #98

Be sure to use only what you have learned to this point (in this book) for solving the word problems below:

1. There are 87 books on two shelves. If there are 49 books on the bottom shelf, how many are on the top shelf?

2. Arlene has 76 cents. Janet has 53 cents. How much more money does Arlene have?

3. The temperature in the desert was 120 degrees in the daytime and 75 degrees at night. How many degrees warmer was it during the day? How many degrees cooler was it during the night?

4. A town collected $1,143,201 in taxes last year and $987,423 this year. How much less did the town collect this year?

5. A new car was being sold for $18,015. A used car was being sold for $9,999. How much more did the new car cost?

6. A city had a population of 3,280,096 on the last census. Since then the city has lost 793,247 persons. What is the present population?

7. Mary paid $73 for a pen and a calculator. The cost of the pen was $17. What was the cost of the calculator?

8. Mrs. Hale traveled by car from Jonestown through Smithtown to Allentown. The entire trip covered 307 miles. If the distance from Smithtown to Allentown is 179 miles, what is the distance from Jonestown to Smithtown?

9. In the morning there were 215 muffins. By afternoon 147 were sold. How many muffins were left?

10. Marcia has $40 and buys a blouse for $23. How much money does she have left?

11. A bus was being driven to a town 300 miles away. The driver took a rest after 117 miles. How many more miles does she still have to drive?

12. In a city's election for mayor 3,146,538 votes were cast for two candidates: Mr. Hill and Ms. Hart. Mr. Hill received 1,472,846 votes. Who won the election? By how many votes did the new mayor win?

13. Henry Simmons paid $3,807 for the purchase of a printing machine. If he paid a tax of $282 on the price of the machine, what was its price?

14. Two trucks are on sale. One costs $28,497 and the other costs $31,000. What is the difference in their prices?

15. A number of pens have the same price. If two of them cost $48, what is the cost of one pen? What is the cost of three pens?

16. A number of calculators have the same price. If eight of them cost $4,862, what is the cost of four calculators? What is the cost of twelve calculators?

17. If five books of the same price cost $185 and the price of one is $37, what is the price of four books? Three books? Two books? Nine Books?

18. If seven rings of the same price cost $4,123 and the cost of one is $589, what is the cost of six rings? Five rings? Four rings? Three rings? Two rings? Twelve rings?

19. A number of machines have the same weight. Twelve of them weigh 888 pounds. Eight of them weigh 592 pounds and three of them weigh 222 pounds. What is the weight of twenty machines?

FINDING THE MISSING NUMBER
FACILITY EXERCISES #99

Write your answers in the spaces provided.

(1)

3 + 2 = ___

5 – 2 = ___

5 – 3 = ___

(2)

3 + ___ = 5

5 – ___ = 3

5 – ___ = 2

(3)

___ + 2 = 5

___ – 2 = 3

___ – 3 = 2

(4)

6 + 1 = ___

7 – 6 = ___

7 – 1 = ___

(5)

6 + ___ = 7

7 – ___ = 1

7 – ___ = 6

(6)

___ + 1 = 7

___ – 6 = 1

___ – 1 = 6

(7)

9 – 4 = ___

5 + 4 = ___

9 – 5 = ___

(8)

9 – ___ = 5

5 + ___ = 9

9 – ___ = 4

(9)

___ – 4 = 5

___ + 4 = 9

___ – 5 = 4

(10)

10 – 4 = ___

6 + 4 = ___

10 – 6 = ___

(11)

10 – ___ = 6

6 + ___ = 10

10 – ___ = 4

(12)

___ – 4 = 6

___ + 4 = 10

___ – 6 = 4

(13)

3 + 4 = ___

3 + ___ = 7

7 − ___ = 4

(14)

7 − 4 = ___

___ + 4 = 7

___ − 4 = 3

(15)

7 − 3 = ___

7 − ___ = 3

___ − 3 = 4

(16)

9 − 6 = ___

6 + 3 = ___

9 − 3 = ___

(17)

___ − 6 = 3

___ + 3 = 9

___ − 3 = 6

(18)

9 − ___ = 3

6 + ___ = 9

9 − ___ = 6

(19)

2 + ___ = 3

3 − ___ = 1

3 − 1 = ___

(20)

___ + 1 = 3

___ − 2 = 1

3 − 2 = ___

(21)

3 − ___ = 2

___ − 1 = 2

2 + 1 = ___

(22)

8 − 3 = ___

8 − 5 = ___

5 + 3 = ___

(23)

8 − ___ = 5

___ − 5 = 3

5 + ___ = 8

(24)

___ − 3 = 5

8 − ___ = 3

___ + 3 = 8

(25)

2 + 4 = ___

___ + 4 = 6

6 − ___ = 4

(26)

6 − 2 = ___

2 + ___ = 6

___ − 4 = 2

(27)

6 − 4 = ___

___ − 2 = 4

6 − ___ = 2

(28)

8 – ___ = 7
1 + ___ = 8
7 + 1 = ___

(29)

7 + ___ = 8
___ – 1 = 7
8 – 1 = ___

(30)

8 – ___ = 1
___ – 7 = 1
8 – 7 = ___

(31)

___ + 8 = 9
___ – 8 = 1
9 – ___ = 1

(32)

___ + 1 = 9
8 + 1 = ___
9 – 1 = ___

(33)

9 – ___ = 8
___ – 1 = 8
9 – 8 = ___

(34)

___ – 1 = 5
6 – ___ = 1
___ + 1 = 6

(35)

___ – 5 = 1
6 – 1 = ___
6 – 5 = ___

(36)

5 + 1 = ___
6 – ___ = 5
___ + 5 = 6

(37)

4 – ___ = 1
___ – 1 = 3
3 + ___ = 4

(38)

4 – ___ = 3
4 – 3 = ___
1 + 3 = ___

(39)

___ – 3 = 1
1 + ___ = 4
4 – 1 = ___

(40)

1 + ___ = 10
10 – ___ = 1
___ – 1 = 9

(41)

9 + ___ = 10
1 + 9 = ___
10 – 9 = ___

(42)

___ – 9 = 1
10 – ___ = 9
10 – 1 = ___

(43)

___ – 4 = 1
4 + ___ = 5
1 + ___ = 5

(44)

___ – 1 = 4
___ + 1 = 5
5 – 1 = ___

(45)

5 – ___ = 4
5 – ___ = 1
5 – 4 = ___

(46)

3 + 3 = ___
6 – 3 = ___

(47)

___ – 3 = 3
6 – ___ = 3

(48)

___ + 3 = 6
3 + ___ = 6

(49)

4 + 4 = ___
8 – 4 = ___

(50)

___ – 4 = 4
4 + 4 = ___

(51)

8 – ___ = 4
___ + 4 = 8

(52)

10 – 5 = ___
5 + 5 = ___

(53)

___ – 5 = 5
___ + 5 = 10

(54)

5 + ___ = 10
10 – ___ = 5

(55)

4 – 2 = ___
2 + ___ = 4

(56)

___ – 2 = 2
4 – ___ = 2

(57)

2 + 2 = ___
___ + 2 = 4

FINDING THE MISSING NUMBER
FACILITY EXERCISES #100

Your teacher will lead you through the examples below:

(1)

___ + 50 = 150

___ – 100 = 50

7 – ___ = 2

(2)

37 – 23 = ___

___ – 200 = 200

84 – ___ = 27

(3)

167 + ___ = 205

167 – ___ = 107

___ + 36 = 70

(4)

789 + 658 = ___

600 – 237 = ___

500 – ___ = 73

(5)

357 – ___ = 57

___ – 209 = 70

400 + ___ = 483

(6)

___ – 48 = 15

145 + ___ = 182

84 – 27 = ___

(7)

84 + 27 = ___

300 + ___ = 417

243 – ___ = 219

(8)

___ + 384 = 679

___ – 384 = 679

679 – ___ = 384

FINDING THE MISSING NUMBER
FACILITY EXERCISES #101

In each example below, use the two numbers shown to find the one which is missing.

(1)

59 – ___ = 32

___ – 27 = 63

41 – 18 = ___

(2)

29 + ___ = 42

___ + 135 = 643

489 + 768 = ___

(3)

40 – ___ = 23

67 + ___ = 81

586 – ___ = 147

(4)

96 – 87 = ___

291 – ___ = 248

400 + ___ = 673

(5)

105 + ___ = 114

___ + 34 = 43

321 – ___ = 123

(6)

8 – ___ = 5

7 + ___ = 10

10 – ___ = 4

(7)

9 + ___ = 16

15 – ___ = 10

___ + 36 = 108

(8)

7 – ___ = 4

6 + 7 = ___

16 + ___ = 25

COLUMN ADDITION
FACILITY EXERCISES #102

Add the following columns.

(1)	(2)	(3)	(4)	(5)
5	6	3	4	8
8	9	2	9	6
+ 3	+ 8	+ 7	+ 7	+ 7

(6)	(7)	(8)	(9)	(10)
4	8	5	9	6
5	4	7	6	5
8	7	3	9	8
+ 6	+ 9	+ 8	+ 8	+ 4

(11)	(12)	(13)	(14)	(15)
4	3	2	8	7
8	8	9	7	3
3	5	8	6	8
2	4	4	7	2
+ 6	+ 7	+ 8	+ 9	+ 5

(16)	(17)	(18)	(19)	(20)
5	6	4	7	9
8	3	2	6	7
4	8	1	3	6
6	1	5	4	8
9	4	8	2	9
+ 4	+ 3	+ 3	+ 5	+ 7

(21)	(22)	(23)	(24)	(25)
7	6	9	5	8
4	2	4	1	9
8	8	9	8	5
6	5	6	4	1
9	7	8	3	4
5	4	9	7	2
3	3	+ 7	5	+ 7
7	+ 8		9	
+ 6			+ 2	

COLUMN ADDITION
FACILITY EXERCISES #103

Add the following:

(1)

```
  4,275
  8,385
  7,025
  6,195
+ 5,115
```

(2)

```
    421
  2,009
 67,084
    275
+ 87,147
```

(3)

```
  573,646
   34,280
+  81,127
```

(4)

```
  4,859
  3,876
  7,695
  9,787
+ 4,936
```

(5)

```
  12,462
   2,955
  27,778
   5,843
+ 31,784
```

(6)

```
   2,739
  30,329
   1,640
  16,458
+  3,665
```

(7)

```
  597
  641
  356
  668
  724
   67
  723
+ 254
```

(8)

```
  484
  603
  765
   80
  106
  567
  474
+ 242
```

(9)

```
  529
  867
  395
  463
  572
  299
  478
+ 656
```

(10)

645
723
947
868
793
528
275
+ 989

(11)

325
842
978
677
593
284
496
+ 888

(12)

674
587
938
868
765
794
869
+ 376

(13)

2,695
3,827
5,674
9,286
+ 7,467

(14)

3,498
2,605
1,380
979
+ 367

(15)

7,846
6,751
8,695
3,879
+ 9,487

(16)

23,498
57,695
+ 48,976

(17)

876,576
295,382
+ 594,768

(18)

679,483
374,958
+ 276,849

FACILITY EXERCISES #104
MIXED PRACTICE

Write your answers in the spaces provided.

1. Find answers to the examples below without use of fingers:

(a) $10 - 6 - 2 + 5 - 3 =$ ___ (c) $7 + 2 - 3 + 1 - 3 - 4 =$ ___

(b) $8 - 2 - 4 - 2 + 6 - 3 + 5 =$ ___ (d) $2 + 3 + 5 - 9 + 7 - 3 =$ ___

Do each addition and subtraction below. Check the answer to each subtraction.

(2)	(3)	(4)
59,483,768 + 74,259,809	74,259,809 − 59,483,768	63,847,290 − 7,479,387

(5)	(6)	(7)
1,847,382 + 68,549	4,070,000 − 153,276	100,010,005 − 725,138

8. Start at zero and count by fours to forty.

9. Start at one and count by fours to forty-one.

10. Start at twenty-three and count backward by threes to two.

11. Read each numeral below:
    (a) 808,008,000 (c) 2,020,000
    (b) 41,414,141,414,141 (d) 5,000,000,050,000

12. Name the number which each digit in 58,479,352,741 represents.

13. Name the value of each place in 58,479,352,741.

14. The number 7,031,574 is in
    (a) which hundreds?
    (b) which ten-thousands?
    (c) which ones?
    (d) which hundred-thousands?
    (e) which millions?
    (f) which tens?
    (g) which thousands?

15. Which is smaller:

(a) 53,292,040 or 53,289,999? Why?

(b) 10,140,001 or 10,138,979? Why?

16. Tell the truth as you do the addition and subtraction examples below:

(a)

$$\begin{array}{r} 730{,}007 \\ -\ 182{,}049 \\ \hline \end{array}$$

(b)

$$\begin{array}{r} 327{,}598 \\ +\ 942{,}562 \\ \hline \end{array}$$

17. Find the missing numbers:

(a) 47 + ____ = 205

(b) 63 − ____ = 16

(c) 287 = ____ − 124

(d) 387 = ____ + 259

18. Add the following:

(a)

$$\begin{array}{r} 486 \\ 897 \\ 426 \\ 735 \\ 398 \\ 679 \\ 965 \\ 878 \\ 697 \\ +\ 789 \\ \hline \end{array}$$

(b)

$$\begin{array}{r} 748 \\ 967 \\ 679 \\ 854 \\ 796 \\ 587 \\ 968 \\ 879 \\ 586 \\ 697 \\ 754 \\ 897 \\ +\ 649 \\ \hline \end{array}$$

Use what you have learned to this point (in this book) for solving the word problems below:

19. If a bar of candy costs \$2 more than a \$5 bag of cookies, how much does the candy bar cost? How much would you pay for a candy bar and a bag of cookies? How much would you pay for 3 candy bars and 4 bags of cookies?

20. If a book costs \$17 more than a \$6 pen, how much would you pay for one book and one pen? For four pens? For five books? For five pens and four books?

21. If three machines cost \$6,584, what is the cost of nine of them?

22. At the stadium 43,894 persons attended the first playoff game, 49,487 attended the second and 51,629 attended the third. What is the total playoff attendance?

23. One car costs \$26,584. Find the cost of six cars at the same price.

24. If eight pens cost \$184 and the cost of one is \$23, what is the cost of seven pens? Six pens? Five pens? Four pens? Three pens? Two pens? Fourteen pens?

25. If you have \$1,729 and you purchase one calculator for \$247, how much money would you have left? How much money would you have left after you purchase two calculators? How many calculators could you purchase altogether?

26. How many printing machines could you purchase out of \$17,256 if one of them costs \$2,876?

27. If you wish to share \$20 among five children, how many times could you give each child \$1? How much money would each child get?

28. If you wish to share 193 apples among 47 persons, how many times would each person get one apple? How many apples would each person get? How many apples are left over?

29. Thirty-four peaches fit into one box. How many boxes would be needed to pack 136 peaches?

30. At the Jamestown Post Office 2,476 letters were mailed on Monday, 3,759 were mailed on Tuesday, 2,009 on Wednesday, 4,987 on Thursday, 2,864 on Friday and 1,694 on Saturday. How many letters were mailed that week at Jamestown Post Office?

31. Mrs. Lane packed 5 boxes full of toys. Each box contained 169 toys and she had 53 of them left over. How many toys did she have altogether?

32. Mr. Crane has 421 cassettes. If each box contains 78 cassettes, how many boxes will he be able to pack completely? How many will he have left over?

33. Sally Simms went shopping with $500. She spent $159 on one item, $238 on another and $99 on a third. How much money will she have left over after these three purchases?

34. Bill has $270. How many chairs could he buy at $34 each? How much money would he have left over?

35. If two pens cost $5 and three caps cost $7, what is the cost of six pens and six caps?

36. Mr. Wilson has $550. How many tables could he buy at $43 each? How much money would he have left over? How many pens at $10 each could he buy with the left-over money?

37. If six sofas cost $2,556 and the price of one is $426, what is the price of five sofas? Four sofas? Three sofas? Ten sofas?

38. A number of handbags cost the same amount. Fourteen of them cost $448. What is the cost of seven of them? If one handbag costs $32 what is the cost of 5 of them? What is the cost of 22 of them?